Heat and Mass Transfer

Series Editors: D. Mewes and F. Mayinger

Harald Mehling · Luisa F. Cabeza

Heat and cold storage with PCM

An up to date introduction into basics and applications

With 208 Figures and 28 Tables

Dr. Harald Mehling
Bayerisches Zentrum für
Angewandte
Energieforschung e.V. (ZAE Bayern)
Abt. Energieumwandlung und
Speicherung
Walther-Meißner-Str. 6
85748 Garching
Germany
mehling@muc.zae-bayern.de

Prof. Dr. Luisa F. Cabeza
Universitat de Lleida
Departament d'Informàtica
i Enginyeria Industrial
c/ Jaume II 69,
25001 Lleida
Spain
lcabeza@diei.udl.es

Series Editors
Prof. Dr.-Ing. Dieter Mewes
Universität Hannover
Institut für Verfahrenstechnik
Callinstr. 36
30167 Hannover, Germany

Prof. em. Dr.-Ing. E.h. Franz Mayinger
Technische Universität München
Lehrstuhl für Thermodynamik
Boltzmannstr. 15
85748 Garching, Germany

ISBN: 978-3-540-68556-2   e-ISBN: 978-3-540-68557-9

DOI 10.1007/978-3-540-68557-9

Heat and Mass Transfer ISSN: 1860-4846

Library of Congress Control Number: 2008929555

© 2008 Springer-Verlag Berlin Heidelberg

This work is subject to copyright. All rights are reserved, whether the whole or part of the material is concerned, specifically the rights of translation, reprinting, reuse of illustrations, recitation, broadcasting, reproduction on microfilm or in any other way, and storage in data banks. Duplication of this publication or parts thereof is permitted only under the provisions of the German Copyright Law of September 9, 1965, in its current version, and permission for use must always be obtained from Springer. Violations are liable to prosecution under the German Copyright Law.

The use of general descriptive names, registered names, trademarks, etc. in this publication does not imply, even in the absence of a specific statement, that such names are exempt from the relevant protective laws and regulations and therefore free for general use.

Cover design: deblik, Berlin, Germany

Printed on acid-free paper

9 8 7 6 5 4 3 2 1

springer.com

Acknowledgements

A book like this cannot be written without the help of many people in one or the other way.

First, the material in this book is based on the work of many people during the past years. They must not be forgotten. Special thanks goes to the companies and those people in R&D who gave the permission to use their graphs, pictures, etc.

To write this book as an introduction and overview, it is necessary to have an open and intense discussion with experts from different areas. In this regard, thanks goes to all the people who have taken part and contributed to the IEA Annex 10 "Phase Change materials and Chemical Reactions for Thermal Energy Storage" and Annex 17 "Advanced Thermal Energy Storage Techniques - Feasibility Studies and Demonstration Projects". Above all, it is necessary to mention Prof. Fredrik Setterwall who served as operating agent in both Annexes. Without the exchange of ideas and knowledge in the different workshops, it would have been impossible to get the background knowledge for writing this book.

Special thanks goes to Stefan Hiebler and Eva Günther for intense and fruitful discussions regarding the content of the book, for their help with some of the material, and the critical reviewing of the manuscript.

Harald Mehling

Preface

The years 2006 and 2007 mark a dramatic change of peoples view regarding climate change and energy consumption. The new IPCC report makes clear that humankind plays a dominant role on climate change due to CO_2 emissions from energy consumption, and that a significant reduction in CO_2 emissions is necessary within decades. At the same time, the supply of fossil energy sources like coal, oil, and natural gas becomes less reliable. In spring 2008, the oil price rose beyond 100 $/barrel for the first time in history. It is commonly accepted today that we have to reduce the use of fossil fuels to cut down the dependency on the supply countries and to reduce CO_2 emissions. The use of renewable energy sources and increased energy efficiency are the main strategies to achieve this goal. In both strategies, heat and cold storage will play an important role.

People use energy in different forms, as heat, as mechanical energy, and as light. With the discovery of fire, humankind was the first time able to supply heat and light when needed. About 2000 years ago, the Romans started to use ceramic tiles to store heat in under floor heating systems. Even when the fire was out, the room stayed warm. Since ancient times, people also know how to cool food with ice as cold storage. Nevertheless, for most of our history, heat and cold storage did not play a significant role for most people in every day life. This has changed during the time of the industrial revolution when the demand for comfort in domestic buildings increased. Today, refrigerators, space heating, and domestic hot water are a part of every household. Thermal energy storage (TES), which is heat and cold storage, plays an important role in many energy systems, not only house holds but also industrial processes. Even though storage itself will never save energy, it is often able to improve a system in a way that it is more energy or cost efficient. The advantage of using heat storage is that it can match supply and demand when they are not at the same time, and second, that a storage can match different powers on demand and supply side. The energy used can have different sources, which are renewable and non-renewable. Especially solar energy is not continuous and thus heat storage is necessary to supply heat reliably. When solar collectors are used to heat domestic hot water, the storage also matches the different powers of the solar collector field, which collects the energy over many hours of the day, to meet the demand of a hot bath that is filled in only several minutes.

The best-known method of thermal energy storage is by changing the temperature of a storage material. Because we can feel the temperature change by our senses, we call this method sensible heat storage. Sensible heat storage is used for example in hot water heat storages or in the floor structure in under floor heating. An alternative method is changing the phase of a material. The best-known examples are ice and snow storage. Their phase change from solid to liquid hereby is especially advantageous, as the melting and solidification occur at a constant temperature, the melting temperature. The storage materials are called phase change materials, or short PCM. Because of the temperature being constant, the heat storage cannot be felt and is called latent heat storage, or short LHS. Due to the constant melting temperature, latent heat storage also allows the stabilization of the temperature. An example is the cooling of drinks using ice. Further on, latent heat storage is also a method of heat storage with a high storage density compared to sensible heat storage when the temperature change in an application is small.

The use of latent heat storage in form of huge ice storages for cooling applications in industry and for space cooling in large buildings is widespread today. Since the oil crisis in the early 1970s other materials than ice with a large range of melting temperatures have been investigated, mainly for solar heating applications. Today, a high variety of storage materials and many products for different applications are available and well established on the market. Examples are the temperature stabilization in transport containers, in peoples clothing, and in buildings. In several other applications, e.g. industrial applications and power generation, the perspectives are improving because of the development in the energy market and new national and international policies.

At this state of the technology of latent heat storage, there is intensive R&D on many aspects as indicated by an increasing number of R&D projects and publications in scientific journals. There were several activities specialized on latent heat storage within the Implementing Agreement (IA) for Energy Conservation through Energy Storage (ECES) of the International Energy Agency (IEA), like IEA Annex 10 "Phase Change materials and Chemical Reactions for Thermal Energy Storage" and Annex 17 "Advanced Thermal Energy Storage Techniques - Feasibility Studies and Demonstration Projects" that finished in 2006. Ongoing or recently started activities are Annex 14 and Annex 20, and Task 32 of the Solar Heating and Cooling program. There are regularly organized conferences where sessions on latent heat storage are included. Examples are the thermal energy storage conferences (called STOCK conferences) organized by the ECES IA of IEA, the last ones being FUTURESTOCK (2003) and ECOSTOCK (2006), and others like EUROSUN (2002 and 2006). A conference dealing with heat transfer fluids with enhanced thermal storage capacity is also organized regularly, the last one being the "7th Conference on Phase Change Materials and Slurries for Refrigeration and Air Conditioning" (2006).

Despite the interest in the technology of PCM by scientists and engineers working in R&D, in industry, at universities, and at institutes, indicated by the large number of publications in journals and presentations at conferences, there is no single source available that can serve as an "introduction into basics and applications". The only book available specialized on latent heat storage was written some 20 years ago by G.A. Lane and has two volumes. The titles are "Solar Heat Storage: Latent Heat Material - Volume I: Background and Scientific Principles" published in 1983 and "Solar Heat Storage: Latent Heat Material - Volume II: Technology" published in 1986. In some aspects, this book is very detailed and not really an introduction, in many other aspects the technology has advanced significantly so that the description in the book is out of date. For example, today companies have taken over the production of PCM and available technologies for their encapsulation have dramatically changed due to the development in plastics in the past decades. In addition, heat transfer, storage concepts, and application examples are not described thoroughly. Further on, the book is out of print. Recently a new book with the title "Thermal energy storage - Systems and applications" written by I. Dinçer and M.A. Rosen was published. It covers many fields of thermal energy storage like sensible heat storage and exergetic analysis. It is however research oriented and written for advanced students in engineering and practicing engineers. Further on, the part on applications of latent heat storages is restricted to industrial applications, mainly using ice storage. Another recent publication is "Thermal energy storage for sustainable energy consumption – fundamentals, case studies and design" published within the NATO Science series II. Mathematics, Physics and Chemistry. It includes several chapters on the technology and applications of latent heat storage written by the authors of this book. These chapters are based on lectures given within a summer school sponsored by the NATO titled "Advanced Study Institute on Thermal Energy Storage for Sustainable Energy Consumption (TESSEC); Fundamentals - Case Studies - Design". The book covers many technologies for heat and cold storage, and consequently the parts on latent heat storage are only a very brief introduction. Therefore, we decided to write this book as an introduction into basics and applications on heat and cold storage with PCM for researchers and for graduate and PhD students in the fields of science and engineering. The chapter on heating and cooling in buildings can also serve architects as an introduction into the new possibilities given by the application of LHS.

The scope of this book is to summarize and explain the most important basics and applications in a single text. To make the book a readable introduction, we tried to keep the length in the range of 300 pages. This means it cannot be a complete coverage or review of PCM technology; the focus is on the explanation of general concepts and their discussion on selected examples. To go beyond an introduction, each chapter supplies many references with focus on such references that serve as an introduction into a special aspect, interesting websites, and other similar information.

As an introductory text, we tried to write the book in a way that things can be understood at the level of a graduate student of science or engineering. Due to the phase change, latent heat storage is considerably more complex than sensible heat storage and the understanding of many technical aspects requires long experience. Therefore, in some cases things have been simplified to allow the reader to understand general concepts. Further on, the book contains many derivations of basic equations, examples, graphs, and tables. To make reading easier, the appendix also includes a list of the most important definitions.

The selection of the material and its order is crucial for an introductory text. This book is based on the experience of several years of presentations and publications for scientists and for non-scientists. Nevertheless, the selection of the material and its order within this book has been reworked many times to give it a logical structure and make it not just a compilation of material. Additionally, the nomenclature used in the past was changed where necessary to be consistent throughout the whole book.

The discussion in this book follows the order from materials to components, then to systems, and finally to applications. The book covers the following topics, each in a single chapter:

1. Basic thermodynamics of thermal energy storage
2. Solid-liquid phase change materials
3. Determination of physical and technical properties
4. Heat transfer basics
5. Design of latent heat storages
6. Integration of active storages into systems
7. Applications in transport and storage containers
8. Applications for the human body
9. Applications for heating and cooling in buildings

Applications in industry and power generation are currently in their first stage, besides ice storage, which is already treated in other books. These applications are therefore not included here. Considerable space has been devoted to measurement of properties. This topic becomes more and more important with increasing commercialization of PCM and until now has often been done without sufficient accuracy.

I hope that this book is helpful to all its readers. In any case, any kind of feedback is appreciated.

Harald Mehling

Contents

1. **Basic thermodynamics of thermal energy storage** 1
   - 1.1 Methods for thermal energy storage 1
     - 1.1.1 Sensible heat 1
     - 1.1.2 Latent heat of solid-liquid phase change 2
     - 1.1.3 Latent heat of liquid-vapor phase change 4
     - 1.1.4 Heat of chemical reactions 5
   - 1.2 Potential applications of latent heat storage with solid-liquid phase change 6
     - 1.2.1 Temperature control 6
     - 1.2.2 Storage of heat or cold with high storage density 7
   - 1.3 References 9

2. **Solid-liquid phase change materials** 11
   - 2.1 Physical, technical, and economic requirements 11
   - 2.2 Classes of materials 13
     - 2.2.1 Overview 13
     - 2.2.2 Detailed discussion 15
   - 2.3 Typical material problems and possible solutions 26
     - 2.3.1 Phase separation solved by mixing, gelling, or thickening 26
     - 2.3.2 Subcooling and methods to reduce it 34
     - 2.3.3 Encapsulation to prevent leakage and improve heat transfer 37
     - 2.3.4 Mechanical stability and thermal conductivity improved by composite materials 39
       - 2.3.4.1 Mechanical stability 39
       - 2.3.4.2 Thermal conductivity 40
   - 2.4 Commercial PCM, PCM composite materials, and encapsulated PCM 41
     - 2.4.1 PCM 42
     - 2.4.2 PCM composite materials 43
       - 2.4.2.1 PCM composite materials to improve handling and applicability 44
       - 2.4.2.2 PCM-graphite composites to increase the thermal conductivity 45
     - 2.4.3 Encapsulated PCM 48
       - 2.4.3.1 Examples of macroencapsulation 49
       - 2.4.3.2 Examples of microencapsulation 51
   - 2.5 References 52

3. **Determination of physical and technical properties** 57
   - 3.1 Definition of material and object properties 57
   - 3.2 Stored heat of materials 59

   3.2.1 Basics of calorimetry 59
   3.2.2 Problems in doing measurements on PCM 64
   3.2.3 Problems in presenting data on PCM 66
   3.2.4 Calorimeter types and working principles 69
      3.2.4.1 Differential scanning calorimetry in dynamic mode 69
      3.2.4.2 Differential scanning calorimetry in steps mode 78
      3.2.4.3 Differential scanning calorimetry with temperature
          modulation (m-DSC) 80
      3.2.4.4 T-History method 80
 3.3 Heat storage and heat release of PCM-objects 84
   3.3.1 Air and other gases as heat transfer medium 85
   3.3.2 Water and other liquids as heat transfer medium 89
      3.3.2.1 Mixing calorimeter 89
      3.3.2.2 Setup derived from power compensated DSC 90
 3.4 Thermal conductivity of materials 91
   3.4.1 Stationary methods 92
   3.4.2 Dynamic methods 93
 3.5 Cycling stability of PCM, PCM-composites, and PCM-objects 95
   3.5.1 Cycling stability with respect to the stored heat 95
   3.5.2 Cycling stability with respect to heat transfer 96
 3.6 Compatibility of PCM with other materials 97
   3.6.1 Corrosion of metals 98
   3.6.2 Migration of components in plastics 101
 3.7 References 102

4 Heat transfer basics 105
 4.1 Analytical models 106
   4.1.1 1-dimensional semi-infinite PCM layer 106
   4.1.2 1-dimensional semi-infinite PCM layer with boundary effects 108
   4.1.3 Cylindrical and spherical geometry 113
   4.1.4 Layer with finite thickness 118
   4.1.5 Summary and conclusion for analytical models 119
 4.2 Numerical models 120
   4.2.1 1-dimensional PCM layer 120
   4.2.2 Inclusion of subcooling using the enthalpy method 126
   4.2.3 Relation between h(T) functions and phase diagrams 128
 4.3 Modellization using commercial software 131
 4.4 Comparison of simulated and experimental results 132
   4.4.1 1-dimensional PCM layer without subcooling 132
   4.4.2 1-dimensional PCM layer with subcooling 133
 4.5 Summary and conclusion 134
 4.6 References 135

5 Design of latent heat storages 137
5.1 Boundary conditions and basic design options 137
5.1.1 Boundary conditions on a storage 137
5.1.2 Basic design options 138
5.2 Overview on storage types 141
5.3 Storages with heat transfer on the storage surface 142
5.3.1 Insulated environment 143
5.3.1.1 Construction principle and typical performance 143
5.3.1.2 Example 143
5.3.1.3 Heat transfer calculation 144
5.3.2 No insulation and good thermal contact between storage and demand 145
5.3.2.1 Construction principle and typical performance 145
5.3.2.2 Example 145
5.3.2.3 Heat transfer calculation 145
5.4 Storages with heat transfer on internal heat transfer surfaces 146
5.4.1 Heat exchanger type 146
5.4.1.1 Construction principle and typical performance 147
5.4.1.2 Example 148
5.4.1.3 Heat transfer calculation 149
5.4.1.4 Further information 158
5.4.2 Direct contact type 158
5.4.2.1 Construction principle and typical performance 159
5.4.2.2 Example 160
5.4.2.3 Heat transfer calculation 161
5.4.2.4 Further information 161
5.4.3 Module type 162
5.4.3.1 Construction principle and typical performance 162
5.4.3.2 Examples 163
5.4.3.3 Heat transfer calculation 164
5.4.3.4 Further information 168
5.5 Storages with heat transfer by exchanging the heat storage medium 168
5.5.1 Slurry type 169
5.5.1.1 Construction principle and typical performance 169
5.5.1.2 Example 170
5.5.1.3 Heat transfer calculation 172
5.5.1.4 Further information 173
5.5.2 Sensible liquid type 174
5.5.2.1 Construction principle and typical performance 174
5.5.2.2 Example 175
5.5.2.3 Heat transfer calculation 176
5.5.2.4 Further information 176
5.6 References 177

6 Integration of active storages into systems ... 181
6.1 Integration goal ... 181
6.2 Integration concepts ... 182
6.2.1 General concepts ... 182
6.2.2 Special examples ... 184
6.3 Cascade storages ... 185
6.4 Simulation and optimization of systems ... 188
6.5 References ... 189

7 Applications in transport and storage containers ... 191
7.1 Basics ... 191
7.1.1 Ideal cooling of an object in ambient air ... 191
7.1.2 Ideal cooling of an insulated object in ambient air ... 193
7.1.3 Ideal cooling of an insulated object with PCM in ambient air ... 195
7.1.4 Real cooling of an insulated object with PCM in ambient air ... 196
7.2 Examples ... 197
7.2.1 Multi purpose transport boxes and containers ... 197
7.2.2 Thermal management system ... 198
7.2.3 Containers for food and beverages ... 199
7.2.4 Medical applications ... 200
7.2.5 Electronic equipment ... 201
7.3 References ... 202

8 Applications for the human body ... 205
8.1 Basics ... 205
8.1.1 Energy balance of the human body ... 205
8.1.2 Potential of PCM ... 206
8.1.3 Methods to apply the PCM ... 207
8.1.3.1 Macroencapsulated PCM ... 207
8.1.3.2 Microencapsulated PCM ... 207
8.1.3.3 Composite materials ... 209
8.2 Examples ... 209
8.2.1 Pocket heater ... 210
8.2.2 Vests for different applications ... 210
8.2.3 Clothes and underwear ... 211
8.2.4 Kidney belt ... 212
8.2.5 Plumeaus and sleeping bags ... 212
8.2.6 Shoe inlets ... 213
8.2.7 Medical applications ... 214
8.3 References ... 214

9 Applications for heating and cooling in buildings ... 217
9.1 Basics of space heating and cooling ... 218
9.1.1 Human comfort requirements ... 218

9.1.2 Heat production, transfer, and storage in buildings........................220
9.1.3 Potential of using PCM 220
    9.1.3.1 Potential of PCM for temperature control 221
    9.1.3.2 Potential of PCM for heat or cold storage with high
            storage density 225
9.1.4 Natural and artificial heat and cold sources 227
    9.1.4.1 Space cooling 227
    9.1.4.2 Space heating 231
9.1.5 Heat transfer...233
    9.1.5.1 Heating or cooling from a surface.......................................233
    9.1.5.2 Heating or cooling by supplying hot or cold air.................234
9.2 Examples for space cooling 234
    9.2.1 Building materials 235
        9.2.1.1 Gypsum plasterboards with microencapsulated paraffin......236
        9.2.1.2 Plaster with microencapsulated paraffin 237
        9.2.1.3 Concrete with microencapsulated paraffin.........................238
        9.2.1.4 Panels with shape-stabilized paraffin.................................240
    9.2.2 Building components 241
        9.2.2.1 Ceiling with PCM 241
        9.2.2.2 Blinds with PCM..243
    9.2.3 Active systems using air as heat transfer fluid 244
        9.2.3.1 Systems integrated into the ceiling.....................................245
        9.2.3.2 Systems integrated into the wall 246
        9.2.3.3 Systems integrated into the floor..247
        9.2.3.4 Decentralized cooling and ventilation unit.........................249
        9.2.3.5 Systems integrated into a ventilation channel 252
    9.2.4 Active building materials and components using a liquid
          heat transfer fluid for heat rejection 254
        9.2.4.1 PCM-plaster with capillary sheets......................................255
        9.2.4.2 Cooling ceiling with PCM-plasterboard 256
    9.2.5 Storages with active heat supply and rejection using a liquid
          heat transfer fluid 256
        9.2.5.1 Heat exchanger and module type storages using
               artificial ice..258
        9.2.5.2 Heat exchanger and module type storages using
               other PCM than ice 263
        9.2.5.3 Direct contact type storage using artificial ice...................263
        9.2.5.4 Storages using natural ice and snow 264
        9.2.5.5 Direct contact systems using other PCM 266
        9.2.5.6 Slurry type storages using artificial ice..............................266
        9.2.5.7 Slurry type storages using other PCM than water / ice 269
    9.2.6 Alternative integration concepts..271
9.3 Examples for space heating..273
    9.3.1 Solar wall 274

        9.3.2 Daylighting element 277
        9.3.3 Floor heating systems 280
                9.3.3.1 Floor heating system with hot water 280
                9.3.3.2 Floor heating system with electrical heating 281
                9.3.3.3 Floor heating system using hot air 281
        9.3.4 Solar air heating and ventilation system 282
        9.3.5 Storage for heating with hot water 284
                9.3.5.1 Heat exchanger type approach 284
                9.3.5.2 Module type approach 286
                9.3.5.3 Direct contact type approach 288
                9.3.5.4 Slurry type approach 289
    9.4 Further information 289
    9.5 References 291

10 Appendix **297**

11 Index **305**

1 Basic thermodynamics of thermal energy storage

In this chapter, different methods of thermal energy storage are first described with respect to their basic characteristics, and then compared with each other. The comparison serves as a basic background to understand what the special advantages and disadvantages of latent heat storage are and when it is more or less useful for thermal energy storage than other methods.

1.1 Methods for thermal energy storage

Thermal energy storage (TES), also commonly called heat and cold storage, allows the storage of heat or cold to be used later. To be able to retrieve the heat or cold after some time, the method of storage needs to be reversible. Fig.1.1 shows some possible methods; they can be divided into physical and chemical processes.

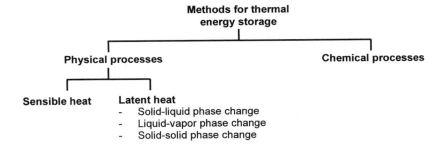

Fig. 1.1. Possible methods of reversible storage of heat and cold.

To understand the distinct advantages of each method, and especially of latent heat storage, it is necessary to get an overview on the different methods of thermal energy storage.

1.1.1 Sensible heat

By far the most common way of thermal energy storage is as sensible heat. As fig.1.2 shows, heat transferred to the storage medium leads to a temperature increase of the storage medium. A sensor can detect this temperature increase and the heat stored is thus called *sensible heat*.

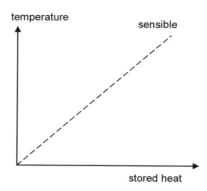

Fig. 1.2. Heat storage as sensible heat leads to a temperature increase when heat is stored.

The ratio of stored heat ΔQ to the temperature rise ΔT is the *heat capacity C* of the storage medium

$$\Delta Q = C \cdot \Delta T = m \cdot c \cdot \Delta T . \tag{1.1}$$

Often the heat capacity is given with respect to the amount of material, the volume, or the mass. It is then called molar, volumetric, or mass *specific heat capacity* and denoted by c. Eq.1.1 shows the case of the mass specific heat capacity were m is the mass of the storage material. Sensible heat storage is often used with solids like stone or brick, or liquids like water, as storage material. Gases have very low volumetric heat capacity and are therefore not used for sensible heat or cold storage.

Sensible heat storage is by far the most common method for heat storage. Hot water heat storages are used for domestic heating and domestic hot water in every household. In recent years, heat storage in the ground has also been applied more and more. As an introduction into the different technologies of sensible heat storage, the interested reader can use the books of Dincer and Rosen 2002, Hadorn 2005, and Paksoy 2007.

1.1.2 Latent heat of solid-liquid phase change

If heat is stored as latent heat, a phase change of the storage material is used. There are several options with distinct advantages and disadvantages. The phase change solid-liquid by melting and solidification can store large amounts of heat or cold, if a suitable material is selected. Melting is characterized by a small volume change, usually less than 10 %. If a container can fit the phase with the larger volume, usually the liquid, the pressure is not changed significantly and consequently melting and solidification of the storage material proceed at a constant temperature. Upon melting, while heat is transferred to the storage material, the material still keeps its temperature constant at the *melting temperature*, also called *phase change temperature* (fig.1.3).

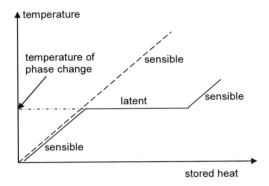

Fig. 1.3. Heat storage as latent heat for the case of solid-liquid phase change.

If the melting is completed, further transfer of heat results again in sensible heat storage. The storage of the heat of melting cannot be detected from the temperature, because the melting proceeds at a constant temperature. The heat supplied upon melting is therefore called *latent heat*, and the process *latent heat storage (LHS)*. Because of the small volume change, the stored heat is equal to the enthalpy difference (section 3.2.1)

$$\Delta Q = \Delta H = m \cdot \Delta h .\tag{1.2}$$

The latent heat, that is the heat stored during the phase change process, is then calculated from the enthalpy difference ΔH between the solid and the liquid phase. In the case of solid-liquid phase change, it is called solid-liquid *phase change enthalpy*, *melting enthalpy*, or *heat of fusion*. Materials with a solid-liquid phase change, which are suitable for heat or cold storage, are commonly referred to as *latent heat storage material* or simply *phase change material (PCM)*.

Some solid-solid phase changes have the same characteristics as solid-liquid phase changes, but usually do not posses a large phase change enthalpy. However, there are exceptions and they are used in a few applications. Further on, even materials with a solid-liquid phase change can be combined with other materials in a way that they do not show the liquid properties any more (section 2.3.4.1). For this reason, whenever talking about phase change in general, the terms phase change temperature T_{pc} and phase change enthalpy $\Delta_{pc}h$ are used, instead of the terms melting temperature and melting enthalpy.

1.1.3 Latent heat of liquid-vapor phase change

The liquid-vapor phase change by evaporation and condensation also usually has a large phase change enthalpy; however, the process of evaporation strongly depends on the boundary conditions:

- In closed systems with constant volume, evaporation leads to a large increase of the vapor pressure. A consequence of the rising vapor pressure is that the temperature necessary for a further phase change also rises. Liquid-vapor phase change in a constant volume is therefore usually not useful for heat storage.
- In closed systems at constant pressure, evaporation leads to a large volume change. This is difficult to realize and thus also not applied for heat storage.
- Open systems at constant, that means ambient pressure, are a third option. This option avoids a change of the phase change temperature. Upon loading the storage with heat, the storage material is evaporated. Because the system is open, the storage material is lost to the environment. To retrieve the stored heat from the storage, the storage material has to be retrieved from the environment. This means it has to be a natural part of the environment. The only technically used material today is water.

If only one component is present, like water, the process is specifically called *homogeneous evaporation-condensation*. Water however does not condense at a high rate from the atmosphere by itself. Therefore, the condensation must be improved by a reduction of the water vapor pressure in the storage using a hygroscopic surface. The hygroscopic effect can be caused by the adsorption of the water at the surface of a solid or the absorption in a liquid (fig.1.4). Due to the second component, this is called *heterogeneous evaporation-condensation*.

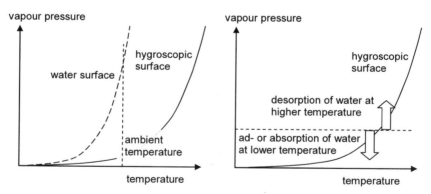

Fig. 1.4. Change of water vapor pressure as a function of temperature using a hygroscopic surface (left) and the application of this effect for heat storage (right).

When water is absorbed in a liquid, usually a salt solution that is strongly hygroscopic at ambient temperature is used. In a first step, the salt solution will absorb water from the atmosphere and the heat of solution and the heat of condensation are released (fig.1.4). While absorbing water the salt solution is diluted. In a second step, the water can be released (desorbed) again by supplying heat to the salt solution and thereby storing the heat in the salt solution. While desorbing water, the concentration of the salt solution rises again to the level at the beginning. The same effect is used in ad- and desorption on solid surfaces. Instead of having a hygroscopic liquid, the water is here adsorbed at the surface of micro pores in a highly porous solid like zeolite.

Because of the additional energetic effect due to the solution of the vapor in the liquid or the adsorption at the surface, the total enthalpy change in a heterogeneous evaporation or condensation is higher than for the corresponding homogeneous evaporation or condensation. The amount of heat stored when using ad- and desorption can be calculated with eq.1.2, using the appropriate enthalpy change. The process of adsorption and desorption, that is the heat release and heat storage proceed at different temperatures, as shown in fig.1.4. This is a significant difference to solid-liquid phase changes.

A good introduction and a list of further references into this topic give Hauer 2007a and Bales et al. 2005. Applications of the technology are described in Hauer 2007b and Hauer et al. 2007.

1.1.4 Heat of chemical reactions

When a chemical reaction takes places, there is a difference between the enthalpy of the substances present at the end of the reaction and the enthalpy of the substances at the start of the reaction. This enthalpy difference is known as *heat of reaction*. If the reaction is endothermic, it will absorb this heat while it takes place; if the reaction is exothermic, it will release this heat. Any chemical reaction with high heat of reaction can be used for thermal energy storage if the products of the reaction can be stored and if the heat stored during the reaction can be released when the reverse reaction takes place. If all components involved in the chemical reaction are solid or liquid the necessary storage space is comparatively small. If any of the components is a gas, the same restrictions apply as for liquid-vapor phase changes. A possible solution is to use the oxidation and reduction of chemicals with oxygen O_2, as oxygen is easily available from the ambient air. The amount of heat stored when using chemical reactions can be calculated with eq.1.2, using the appropriate enthalpy change. As the binding energy in a chemical reaction is usually large, the temperature necessary to destroy the bond is usually high.

Good introductions and a list of further references into the topic give Kato 2007 and van Berkel 2005. Kato 2004 describes several examples of chemical reactions for high temperature heat storage.

1.2 Potential applications of latent heat storage with solid-liquid phase change

In general, the term "latent heat" describes the heat of solid-solid, solid-liquid, and liquid-vapor phase changes. However, the terms "latent heat storage" and "phase change material" are commonly only used for the first two kinds of phase changes, and not for liquid-vapor phase changes. In a liquid-vapor phase change, the phase change temperature strongly depends on the boundary conditions, and therefore the phase change is not just used for storage of heat alone. Usually it is connected with a pressure and a temperature difference between charging and discharging. In this sense, for the rest of the book "latent heat" and "phase change material" will refer to solid-liquid phase changes only.

Potential fields of application of PCM can be found directly from the basic difference between sensible and latent heat storage, shown in fig.1.5.

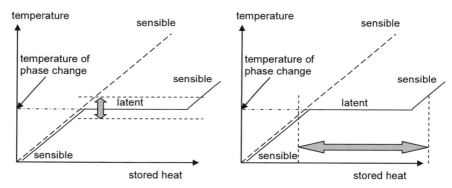

Fig. 1.5. Potential fields of application of PCM: temperature control (left) and storage and supply of heat or cold with small temperature change (right).

They can be divided as a simplification into temperature control and storage and supply of heat or cold with small temperature change.

1.2.1 Temperature control

As shown on the left of fig.1.5, heat can be supplied to or extracted from a phase change material without a significant change of its temperature. Therefore, PCM can be used to stabilize the temperature in an application. Examples are the use of ice to keep drinks and food cold, the stabilization of the indoor temperature in a building, or the temperature in transport boxes.

1.2.2 Storage of heat or cold with high storage density

From another point of view, shown on the right in fig.1.5, PCM are also able to store large amounts of heat or cold at comparatively small temperature change. A comparison of energy storage densities achieved with different methods shows tab.1.1.

Table 1.1. Comparison of typical storage densities of different energy storage methods (3.6 MJ = 1 kWh).

	MJ/m^3	kJ/kg	Comment
Sensible heat			
granite	50	17	$\Delta T = 20\ °C$
water	84	84	$\Delta T = 20\ °C$
Latent heat of melting			
water	306	330	melting temperature 0 °C
paraffins	180	200	melting temperatures 5 °C – 130 °C
salt hydrates	300	200	melting temperatures 5 °C – 130 °C
salts	600 - 1500	300 - 700	melting temperatures 300 °C – 800 °C
Latent heat of evaporation			
water	2452	2450	ambient conditions
Heat of chemical reaction			
H_2 gas (oxidation)	11	120000	300 K, 1 bar
H_2 gas (oxidation)	2160	120000	300 K, 200 bar
H_2 liquid (oxidation)	8400	120000	20 K, 1 bar
fossil gas	32	-	300 K, 1 bar (Diekmann et al. 1997)
gasoline (petroleum)	33000	43200	(Diekmann et al. 1997)
Electrical energy			
zinc/manganese oxide battery-		180	(Diekmann et al. 1997)
lead battery	-	70 - 180	(Diekmann et al. 1997)

As tab.1.1 shows, PCM can store about 3 to 4 times more heat per volume than is stored as sensible heat in solids or liquids in a temperature interval of 20 °C. This can be a significant advantage in many applications like in domestic space heating. Chemical energy storage in petroleum however shows a storage density about 100 times larger than that of PCM. Because nowadays a big part of the energy is delivered to the end user as electricity, a comparison with the storage of electrical energy is also interesting. Batteries show smaller or comparable storage densities as in latent heat storage. If the final demand is for heat, it is therefore often an option to store heat in a PCM instead of storing electrical energy and converting the electrical energy to heat on demand. If the final demand is cold, the situation is the same.

The large differences in the storage density of up to four orders of magnitude can be understood from a molecular point of view. On a molecular level, each storage method uses different effects with typical storage density. An overview shows tab.1.2.

Table 1.2. Typical molar heat stored using different effects (based on data from Alefeld 1977).

Theoretical model	molar heat stored	storage medium	comments
Sensible heat			
Dulong-Petit	$\Delta Q_{mol} = n \cdot 3 \cdot R \cdot \Delta T$	solids	n = number of atoms per molecule
Dulong-Petit	$\Delta Q_{mol} = n \cdot 1.5 \cdot R \cdot \Delta T$	ideal gases at const. volume	
	$\Delta Q_{mol} = n \cdot 2.5 \cdot R \cdot \Delta T$	ideal gases at const. pressure	
Latent heat of melting-solidification			
Richardson	$1 \cdot R \cdot T_{mp} < \Delta Q_{mol} < 1.5 \cdot R \cdot T_{mp}$	metals	T_{mp} = melting point
Richardson	$\Delta Q_{mol} < 3.0 \cdot R \cdot T_{mp}$	semiconductors	
Richardson	$\Delta Q_{mol} < 4.5 \cdot R \cdot T_{mp}$	salts	
Latent heat of evaporation-condensation			
Trouton	$9 \cdot R \cdot T_{bp} < \Delta Q_{mol} < 13 \cdot R \cdot T_{bp}$	homogeneous evap.-cond. 1 bar (13 applies for water)	T_{bp} = boiling point
Trouton	$15 \cdot R \cdot T_{bp} < \Delta Q_{mol} < 18 \cdot R \cdot T_{bp}$	heterogeneous evap.-cond. 1 bar	
Heat of chemical reactions			
	$50 \cdot R \cdot T_{amb} < \Delta Q_{mol} < 220 \cdot R \cdot T_{amb}$	single chemical bond	T_{amb} = ambient temperature of approx. 20 °C
	$70 \cdot R \cdot T_{amb} < \Delta Q_{mol} < 290 \cdot R \cdot T_{amb}$	double chemical bond	
	$370 \cdot R \cdot T_{amb} < \Delta Q_{mol} < 430 \cdot R \cdot T_{amb}$	triple chemical bond	

The storage of sensible heat uses movements of atoms and molecules and is described by the rule of Dulong-Petit for solids and for ideal gases. For liquids, the description is much more complex. For the case of heat storage using the latent heat of melting and solidification associated with a solid-liquid phase change, the rule of Richardson can be used. The latent heat is proportional to the phase change temperature, which is in this case the melting temperature. The proportionality constant includes a factor which depends on the class of the material and the universal gas constant R (R = 8.3 kJ/kmol·K). As an example, the latent heat of NaCl with a melting temperature of 800 °C according to the rule of Richardson is less than 4.5 · 8.3 kJ/kmol·K · 1173 K ≈ 44000 kJ/kmol. The molar mass is 58 g/mol (23 g/mol for Na and 35 g/mol for Cl). Then, the latent heat of NaCl according to the rule of Richardson should be less than 755 kJ/kg. This is in accordance with the value of 492 kJ/kg given in tab.2.3. The basic principle behind the rule of Richardson is that associated with the phase change from solid to liquid the structure of the material changes and thus also its entropy. Because of the constant phase change temperature, the heat of melting is directly proportional to the entropy change ΔS during phase change

$$\Delta Q = T \cdot \Delta S .$$ (1.3)

For the case of heat storage using the latent heat of evaporation and condensation associated with a liquid-vapor phase change, the rule of Trouton can be used. Here the proportionality constant is larger and the phase change temperature is the boiling temperature. Finally, also chemical reactions can be described in the same manner. Here the proportionality constant is even larger; instead of the phase change temperature, the heat of the chemical reaction is given with respect to ambient temperature.

1.3 References

[Alefeld 1977] Alefeld G.: Basic physical and chemical processes for storage of heat. Proc. of the symposium on "Electrode materials and processes for energy conversion and storage", May 1977, Philadelphia (1977)

[Bales et al. 2005] Bales C., Gantenbein P., Hauer A., Jaehnig D., Kerskes H., Henning H.-M., Nuñez T., Visscher K., Laevemann E., Peltzer M.: Sorption and thermo-chemical storage. In: Hadorn J.-C. (ed.) Thermal energy storage for solar and low energy buildings, pp. 117-155. Servei de Publicacions (UDL), Lleida (2005) ISBN 84-8409-877-X

[Diekmann et al. 1997] Diekmann B., Heinloth K.: Energie – Physikalische Grundlagen ihrer Erzeugung, Umwandlung und Nutzung. Teubner Studienbücher (1997)

[Dincer and Rosen 2002] Dincer I., Rosen M.A.: Thermal energy storage - Systems and applications. John Wiley & Sons, Chichester (2002)

[Hadorn 2005] Hadorn J.-C. (ed.) Thermal energy storage for solar and low energy buildings. Servei de Publicacions (UDL), Lleida (2005) ISBN 84-8409-877-X

[Hauer 2007a] Hauer A.: Sorption theory for thermal energy storage. In: Paksoy H.Ö. (ed.): Thermal energy storage for sustainable energy consumption – fundamentals, case studies and design, pp. 393-408. Springer, (2007), NATO Science series II. Mathematics, Physics and Chemistry – Vol. 234, ISBN 978–1–4020–5289–7

[Hauer 2007b] Hauer A.: Adsorption systems for TES – Design and demonstration projects. In: Paksoy H.Ö. (ed.): Thermal energy storage for sustainable energy consumption – fundamentals, case studies and design, pp. 409-428. Springer, (2007), NATO Science series II. Mathematics, Physics and Chemistry – Vol. 234, ISBN 978–1–4020–5289–7

[Hauer et al. 2007] Hauer A., Lävemann E.: Open absorption systems for air conditioning and thermal energy storage. In: Paksoy H.Ö. (ed.): Thermal energy storage for sustainable energy consumption – fundamentals, case studies and design, pp. 429-444. Springer, (2007), NATO Science series II. Mathematics, Physics and Chemistry – Vol. 234, ISBN 978–1–4020–5289–7

[Paksoy 2007] Paksoy H.Ö. (ed.): Thermal energy storage for sustainable energy consumption – fundamentals, case studies and design. Springer, (2007), NATO Science series II. Mathematics, Physics and Chemistry – Vol. 234, ISBN 978–1–4020–5289–7

[Kato 2004] Kato Y.: Chemical thermal energy storage system using metal oxide reactions for high-temperature heat utilization. Presented at 6th Workshop of IEA ECES Annex 17 "Advanced Thermal Energy Storage through Phase Change Materials and Chemical Reactions – Feasibility Studies and Demonstration Projects", Arvika, Sweden, 2004. www.fskab.com/Annex17

[Kato 2007] Kato Y.: Chemical energy conversion technologies for efficient energy use. In: Paksoy H.Ö. (ed.): Thermal energy storage for sustainable energy consumption – fundamentals, case studies and design, pp. 377-392. Springer, (2007), NATO Science series II. Mathematics, Physics and Chemistry – Vol. 234, ISBN 978–1–4020–5289–7

[van Berkel 2005] van Berkel J.: Storage of solar energy in chemical reactions. In: Hadorn J.-C. (ed.) Thermal energy storage for solar and low energy buildings, pp. 157-167. Servei de Publicacions (UDL), Lleida (2005) ISBN 84-8409-877-X

2 Solid-liquid phase change materials

In this chapter, basic know-how on phase change materials is summarized. The chapter starts with a description of the basic requirements on a material to use it as phase change material. Then different material classes are discussed with respect to their most important properties, advantages, and disadvantages. Then examples for materials from each material class are given. Usually, a material is not able to fulfill all the requirements. Therefore, for several common material problems solutions are given. The chapter ends with a set of examples showing the currently available range of commercial products.

2.1 Physical, technical, and economic requirements

A suitable phase change temperature and a large melting enthalpy are two obvious requirements on a phase change material. They have to be fulfilled in order to store and release heat at all. However, there are more requirements for most, but not all applications. These requirements can be grouped into physical, technical, and economic requirements.

- Physical requirements, regarding the storage and release of heat:
  - Suitable phase change temperature T_{pc} $\Rightarrow$ to assure storage and release of heat in an application with given temperatures for heat source and heat sink.
  - Large phase change enthalpy $\Delta_{pc}h$ $\Rightarrow$ to achieve high storage density compared to sensible heat storage.
  - Reproducible phase change, also called *cycling stability* $\Rightarrow$ to use the storage material as many times for storage and release of heat as required by an application.

    The number of cycles varies from only one, when the PCM is used for heat protection in the case of a fire, to several thousand cycles when used for heating or cooling of buildings. One of the main problems of cycling stability is phase separation. When a PCM consists of several components, phases with different compositions can form upon cycling. *Phase separation* is the effect that phases with different composition are separated from each other macroscopically. The phases with a composition different from the correct initial composition optimized for heat storage then show a significantly lower capacity to store heat.
  - Little subcooling $\Rightarrow$ to assure that melting and solidification can proceed in a narrow temperature range.

Subcooling (also called *supercooling*) is the effect that a temperature significantly below the melting temperature has to be reached, until a material begins to solidify and release heat (fig.2.1). If that temperature is not reached, the PCM will not solidify at all and thus only store sensible heat.

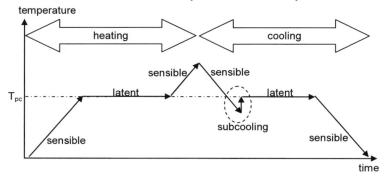

Fig. 2.1. Schematic temperature change during heating (melting) and cooling (solidification) of a PCM with subcooling.

- Good thermal conductivity ⇒ to be able to store or release the latent heat in a given volume of the storage material in a short time, that is with sufficient heating or cooling power.

  If a good thermal conductivity is necessary strongly depends on the application and the design of the storage.

- Technical requirements, regarding the construction of a storage:
  - Low vapor pressure ⇒ to reduce requirements of mechanical stability and tightness on a vessel containing the PCM
  - Small volume change ⇒ to reduce requirements of mechanical stability on a vessel containing the PCM
  - Chemical stability of the PCM ⇒ to assure long lifetime of the PCM if it is exposed to higher temperatures, radiation, gases, …
  - Compatibility of the PCM with other materials ⇒ to assure long lifetime of the vessel that contains the PCM, and of the surrounding materials in the case of leakage of the PCM

    This includes destructive effects as for example the corrosivity of the PCM with respect to other materials, but also other effects that significantly reduce or stop important functions of another material.
  - Safety constraints ⇒ the construction of a storage can be restricted by laws that require the use of non-toxic, non-flammable materials. Other environmental and safety consideration can apply additionally.

- Economic requirements, regarding the development of a marketable product:
  - Low price ⇒ to be competitive with other options for heat and cold storage, and to be competitive with methods of heat and cold supply without storage at all
  - Good recyclability ⇒ for environmental and economic reasons

Lane et al. 1983 gives a detailed discussion of these selection criteria and examples of how to select candidate materials. A first selection of a material is usually done with respect to the physical requirements phase change temperature, enthalpy, cycling stability, and subcooling. The classes of the materials to choose from, as well as lists of examples for different classes with data on phase change temperature and enthalpy, are discussed in the following section.

Usually, a material is not able to fulfill all the requirements mentioned above. For example, the thermal conductivity of a PCM is usually small, inorganic PCM often show subcooling, and compatibility with the container material is not always given. Therefore, different strategies have been developed to cope with these problems. These strategies are discussed in section 2.3. Section 2.4 then concludes with the commercial examples.

2.2 Classes of materials

2.2.1 Overview

Because the two most important criteria, the melting temperature and the melting enthalpy, depend on molecular effects, it is not surprising that materials within a material class behave similar. Fig.2.2 shows the typical range of melting enthalpy and melting temperature of common material classes used as PCM.

14   2 Solid-liquid phase change materials

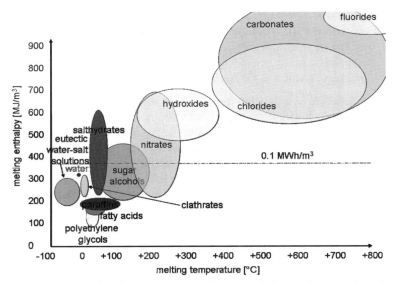

Fig. 2.2. Classes of materials that can be used as PCM and their typical range of melting temperature and melting enthalpy (picture: ZAE Bayern).

By far the best-known PCM is water. It has been used for cold storage for more than 2000 years. Today, cold storage with ice is state of the art and even cooling with natural ice and snow is used again. For temperatures below 0 °C, usually water-salt solutions with a eutectic composition are used. Several material classes cover the temperature range from 0 °C to about 130 °C. Paraffins, fatty acids, and sugar alcohols are organic materials. Salt hydrates are salts with a large and defined amount of crystal water. Clathrates are crystalline structures in which molecules of one type are enclosed in the crystal lattice of another. When the enclosed molecule is from a gas and the surrounding crystal structure is water, the clathrate is also called a gas hydrate. They cover a temperature range from about 0 °C to 30 °C. At temperatures above 150 °C, different salts and their mixtures can be applied.

A close look at fig.2.2 indicates that the energy density is roughly proportional to the melting temperature in K. This can be understood from thermodynamics according to the theory of Richards (tab.1.2). The theory of Richards shows that the melting enthalpy per volume is proportional to the melting temperature, the number of bonds per molecule, and the density divided by the molar mass that relates to the packing density of the molecules or atoms (Lindner 1984).

2.2.2 Detailed discussion

Probably thousands of single materials and mixtures of two or more materials have been investigated for their use as PCM in the past decades. Because the scope of this book is to give an introduction, important and typical examples are discussed here; no attempt is made to give a large or even complete list of materials. For more comprehensive lists of materials the reader should look at the early publications of Steiner et al. 1980, Abhat 1983, Lane 1983 and 1986, Schröder 1985, and more recent publications like Kakiuchi et al. 1998, Hiebler and Mehling 2001, Zalba et al. 2003, Sharma et al. 2004, Farid et al. 2004, and Kenisarin and Mahkamov 2007.

Inorganic materials cover a wide temperature range. Compared to organic materials, inorganic materials usually have similar melting enthalpies per mass, but higher ones per volume due to their high density (fig.2.2). Their main disadvantage is material compatibility with metals, since severe corrosion can be developed in some PCM-metal combinations.

Eutectic water-salt solutions have melting temperatures below 0 °C, because the addition of the salt reduces the melting temperature, and usually good storage density. Tab.2.1 shows a selection of typical examples.

Table 2.1. Examples of eutectic water-salt solutions that have been investigated as PCM (Schröder 1985).

Material	Melting temperature (°C)	Melting enthalpy (kJ/kg)	Thermal conductivity (W/m·K)	Density (kg/m^3)
Al(NO$_3$)$_3$ (30.5 wt.%) / H$_2$O	-30.6	131	-	1283 (liquid)
			-	1251 (solid)
NaCl (22.4 wt.%) / H$_2$O	-21.2	222	-	1165 (liquid)
			-	1108 (solid)
KCl (19.5 wt.%) / H$_2$O	-10.7	283	-	1126 (liquid)
			-	1105 (solid)
H$_2$O	0	333	0.6 (liquid, 20 °C)	998 (liquid, 20°C)
			2.2 (solid)	917 (solid, 0 °C)

Water-salt solutions consist of two components, water and salt, which means phase separation could be a problem. To prevent phase separation, and to achieve a good cycling stability, eutectic compositions are used. *Eutectic compositions* are mixtures of two or more constituents, which solidify simultaneously out of the liquid at a minimum freezing point. Therefore, none of the phases can sink down due to a different density. Further on, eutectic compositions show a melting temperature and good storage density. A more detailed discussion on phase separation and how to get rid of it follows in section 2.3.1. The thermal conductivity of eutectic water-salt solutions is similar to that of water and they can subcool like water by several K or more. The vapor pressure at temperatures as low as their melting temperature is quite small, however like water they can show considerable volume change in the order of 5 to 10 vol.% during melting and solidification. Water can cause severe damage upon freezing and melting for example like cracking stones in winter. Water-salt solutions are chemically very stable, but can cause corrosion to other materials like metals. Compared to water, the addition of a salt usually makes the problem worse. Most of the salt solutions are rather safe, but should not leak in larger amounts. They are usually cheap, often less than 1 €/kg, and therefore the basis for many commercial PCM used in large-scale applications.

The temperature range between 5 °C and 130 °C is covered by salt hydrates. *Salt hydrates* consist of a salt and water in a discrete mixing ratio. It is usually an integral number of water molecules per ion pair of the salt, where a stable crystal structure forms. The bonds are usually ion-dipole bonds or hydrogen bonds. The water molecules are located and oriented in the structure in a well-defined manner. In some structures, the water is more closely oriented to the anion, in others to the cation of the salt.

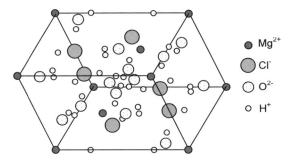

Fig. 2.3. Molecular structure of the salt hydrate $MgCl_2 \cdot 6H_2O$.

An example, magnesium chloride hexahydrate $MgCl_2 \cdot 6H_2O$, is shown in fig.2.3. The lattice consists of two parts: Cl^- ions, and 6 water molecules oriented octahedral around a magnesium ion Mg^{2+} and bound by ion-dipole bonds. Because of the stable crystal structure of salt hydrates, the melting temperature is higher than for water. Tab.2.2 shows a selection of typical examples that are the basis for many commercial PCM. Where several data for the same property are given in the following tables, these data reflect the typical variation of literature data. In some examples, a range is given for the melting temperature, rather than a single value. The different reasons why a material shows a melting range are explained in section 2.3.1.

Table 2.2. Examples of salt hydrates that have been investigated as PCM

Material	Melting temperature (°C)	Melting enthalpy (kJ/kg)	Thermal conductivity (W/m·K)	Density (kg/m³)
$LiClO_3 \cdot 3H_2O$	8	155	-	1530 (liquid)
			-	1720 (solid)
$KF \cdot 3H_2O$	18.5	231	-	1447 (liquid, 20 °C)
			-	1455 (solid, 18 °C)
$CaCl_2 \cdot 6H_2O$	29, 30	171, 190	0.540 (liquid, 39 °C)	1562 (liquid, 32 °C)
			1.088 (solid, 23 °C)	1710 (solid, 25 °C)
$LiNO_3 \cdot 3H_2O$	30	296	-	-
			-	-
$Na_2SO_4 \cdot 10H_2O$	32	254	-	-
			0.554	1485 (solid)
$Na_2HPO_4 \cdot 12H_2O$	35-44	280	0.476 (liquid)	1442 (liquid)
			0.514 (solid)	1522 (solid)
$Na_2S_2O_3 \cdot 5H_2O$	48-55	187, 209	-	1670 (liquid)
			-	1750 (solid)
$Na(CH_3COO) \cdot 3H_2O$	58	226, 264	-	1280 (liquid)
			-	1450 (solid)
$Ba(OH)_2 \cdot 8H_2O$	78	265, 280	0.653 (liquid, 86 °C)	1937 (liquid, 84 °C)
			1.255 (solid, 23 °C)	2180 (solid)
$Mg(NO_3)_2 \cdot 6H_2O$	89, 90	149, 163	0.490 (liquid, 95 °C)	1550 (liquid, 94 °C)
			0.669 (solid, 56 °C)	1636 (solid, 25 °C)
$MgCl_2 \cdot 6H_2O$	117	165, 169	0.570 (liquid, 120 °C)	1450 (liquid, 120 °C)
			0.704 (solid, 110 °C)	1569 (solid, 20 °C)

Salt hydrates often have comparatively high storage density with respect to mass, but even more with respect to volume due to their high density. Because salt hydrates consist of several components, at least one salt and water, they can potentially separate into different phases and thus show problems with cycling stability. In fact, phase separation is a common problem with salt hydrates. Their thermal conductivity is similar to that of water and eutectic water-salt solutions. Most salt hydrates subcool, some of them by as much as 80 K. Their vapor pressure is somewhat lower than for water, as the salt usually reduces the vapor pressure. Salt hydrates melting close to or above 100 °C however show already considerable vapor pressure when melting. The volume change of salt hydrates is up to 10 vol.%. In most, but not all cases, salt hydrates are chemically very stable. However, many of them are potentially corrosive to metals. Regarding the safety of salt hydrates there is a high variation, so the respective data sheets have to be checked. The price of salt hydrates is usually low, in the order of 1 to 3 €/kg unless bought in a pure form.

Above 150 °C, different *salts* can be used as PCM. Tab.2.3 shows a selection of typical examples.

Table 2.3. Examples of salts that have been investigated as PCM

Material	Melting temperature (°C)	Melting enthalpy (kJ/kg)	Thermal conductivity (W/m·K)	Density (kg/m^3)
LiNO$_3$	254	360	0.58 (liquid)	1780 (liquid)
			1.37 (solid)	2140 (solid)
NaNO$_3$	307	172	0.51 (liquid)	1900 (liquid)
			0.59 (solid)	2260 (solid)
KNO$_3$	333	266	0.50 (liquid)	1890 (liquid)
			-	1900 (solid)
MgCl$_2$	714	452	-	2140
			-	-
NaCl	800	492	-	2160
			-	-
Na$_2$CO$_3$	854	276	-	2533
			-	-
KF	857	452	-	2370
			-	-
K$_2$CO$_3$	897	236	-	2290
			-	-

Because the melting enthalpy rises roughly proportional to the melting temperature given in K, salts with high melting temperatures often show a very high melting enthalpy. Fig.2.4 shows an example of a salt: MgCl$_2$. A salt always consists of two components, so theoretically phase separation is a potential problem. However, unless the rare case that two different salt compositions exist, phase separation is not possible.

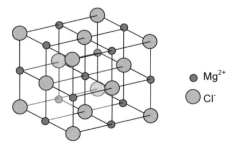

Fig. 2.4. Crystal structure of the salt MgCl$_2$.

The thermal conductivity of salts can be quite good. Subcooling, as far as data are available, is not more than a few K, and their vapor pressure is very low. The volume change from solid to liquid can be up to 10 vol.%. Many of the salts are chemically stable; however, carbonates and nitrates can decompose under unsuitable conditions. Regarding the compatibility to other materials, salts can be corrosive to metals. Their safety differs strongly between different salts. The same holds for their price.

In order to get materials with different melting temperature or improved properties, *mixtures of inorganic materials* have been tested. For example, small amounts of NaCl and KCl are added to $CaCl_2 \cdot 6H_2O$ to achieve a better melting behavior without significant change of the melting temperature (Lane 1986). In another example, the combination of $Mg(NO_3)_2 \cdot 6H_2O$ and $MgCl_2 \cdot 6H_2O$ results in a much lower melting temperature of the mixture compared to the base materials of the mixture. For example, Nagano et al. 2000 systematically investigated the lowering of the melting temperature of $Mn(NO_3)_2 \cdot 6H_2O$ by the addition of other materials. Neuschütz 1999 presented an eutectic mixture of $LiNO_3/Mg(NO_3)_2/H_2O$. Tab.2.4 shows some examples of mixtures (salt – salt hydrate, salt hydrate – salt hydrate, salt – salt) that are based on materials from tab.2.2 and tab.2.3.

Table 2.4. Examples of inorganic mixtures that have been investigated as PCM

Material	Melting temperature (°C)	Melting enthalpy (kJ/kg)	Thermal conductivity (W/m·K)	Density (kg/m³)
4.3 % NaCl + 0.4 % KCl + 48 % $CaCl_2$ + 47.3 % H_2O	27	188	- -	1530 (liquid) 1640 (solid)
58.7 % $Mg(NO_3)_2 \cdot 6H_2O$ + 41.3 % $MgCl_2 \cdot 6H_2O$	58, 59	132	0.510 (liquid, 65 °C) 0.678 (solid, 53 °C)	1550 (liquid, 50 °C) 1630 (solid, 24 °C)
67 % KNO_3 + 33 % $LiNO_3$	133	170	-	-
54 % KNO_3 + 46 % $NaNO_3$	222	100	-	1950 (liquid) 2050 (solid)

The search for mixtures of different materials is usually done experimentally. That means, first mixtures with different compositions are prepared, and then their melting temperature and enthalpy are determined. If positive results are achieved with one or several of the compositions, the composition is optimized in a second set of experiments. This procedure involves a lot of work and is very time consuming. A way to reduce the experimental effort is to calculate the liquidus curves of salt hydrates and multi-component systems (Voigt 1993). Then, only few experiments have to be performed, but such calculations are up to experts in the field.

Besides the material classes of inorganic materials, there are also organic materials used as PCM like paraffins, fatty acids and sugar alcohols. These material classes cover the temperature range between 0 °C and about 200 °C. Due to the covalent bonds in organic materials, most of them are not stable to higher temperatures. In most cases, the density of organic PCM is less than 10^3 kg/m^3, and thus smaller than the density of most inorganic materials like water and salt hydrates. The result is that, with the exception of sugar alcohols, organic materials usually have smaller melting enthalpies per volume than inorganic materials.

The most commonly used organic PCM are paraffins. *Paraffin* is a technical name for an alcane, but often it is specifically used for linear alkanes with the general formula C_nH_{2n+2}, as shown in fig.2.5.

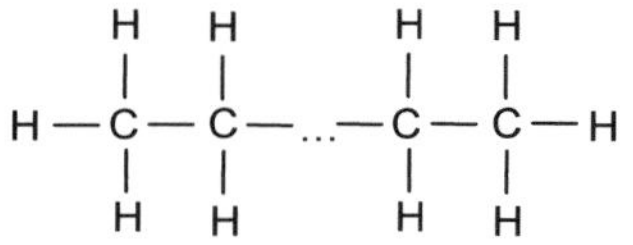

$$\begin{array}{c} H\ \ H\ \ \ \ \ \ \ H\ \ H \\ |\ \ \ | \ \ \ \ \ \ \ |\ \ \ | \\ H-C-C-\ldots-C-C-H \\ |\ \ \ | \ \ \ \ \ \ \ |\ \ \ | \\ H\ \ H\ \ \ \ \ \ \ H\ \ H \end{array}$$

Fig. 2.5. Chemical structure of linear alcanes.

Little information is available on their crystal structure. The picture in fig.2.6 shows the crystal structure of n-hexane. It is based on data published by Boese et al. 1999.

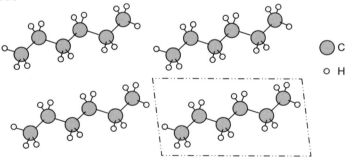

Fig. 2.6. Crystal structure of n-hexane.

Paraffins show good storage density with respect to mass, and melt and solidify congruently with little or no subcooling. Their thermal conductivity is however comparatively low. Regarding the stability of a container, their vapor pressure is usually not significant. Their volume increase upon melting is in the order of 10 vol.%; this is similar to that of many inorganic materials, but less critical as paraffins are softer and therefore build up smaller forces upon expansion. Paraffins are insoluble in water as they are water repellent. They do not react with most

2.2 Classes of materials

common chemical reagents; in fact, the name "paraffin" originates in the Latin and means that they are little reactive. At elevated temperatures, paraffin bonds can crack and the resulting short chain molecules evaporate. Paraffins are combustible and people often conclude that paraffins burn easily. The fact that candles do not burn as a whole shows that this is not correct. The compatibility of paraffins with metals is very good; with plastics however, paraffins can cause softening of the plastic. Paraffins have very few safety constraints. Tab.2.5 shows examples of paraffins. It indicates that with rising number of C atoms in C_nH_{2n+2} the melting temperature increases. The limit is Polyethylene with thousands of C atoms. At very low numbers of C atoms, methane C_1H_4, ethane C_2H_6, and propane C_3H_8 are gasses at ambient conditions. Tetradecane is the smallest n-alcane that melts above 0 °C.

Table 2.5. Examples of paraffins that have been investigated as PCM.

Material	Melting temperature (°C)	Melting enthalpy (kJ/kg)	Thermal conductivity (W/m·K)	Density (kg/m^3)
n-Tetradecane $C_{14}H_{30}$	6	230	- 0.21 (solid)	760 (liquid, 20 °C) -
n-Pentadecane $C_{15}H_{32}$	10	212	- -	770 (liquid, 20 °C) -
n-Hexadecane $C_{16}H_{34}$	18	210, 238	- 0.21 (solid)	760 (liquid, 20 °C) -
n-Heptadecane $C_{17}H_{36}$	19	240	- -	776 (liquid, 20 °C) -
n-Octadecane $C_{18}H_{38}$	28	200, 245	0.148 (liquid, 40 °C) 0.358 (solid, 25 °C)	774 (liquid, 70 °C) 814 (solid, 20 °C)
n-Eicosane $C_{20}H_{42}$	38	283	- -	779 -
n-Triacontane $C_{30}H_{62}$	66	-	- -	775 -
n-Tetracontane $C_{40}H_{82}$	82	-	- -	- -
n-Pentacontane $C_{50}H_{102}$	95	-	- -	779 -
Polyethylene	110-135	200	-	-
C_nH_{2n+2} n up to 100000			-	870-940 (solid, 20 °C)

Pure alkanes are rather expensive. Commercial paraffin is usually obtained from petroleum distillation and contains a number of different hydrocarbons. These mixtures show a melting range and a lower heat of fusion than the pure alkanes.

Alkanes of different chain length are also mixed intentionally, especially to get PCM with different melting temperatures. He et al. 1998 and He et al. 1999 for example investigated binary mixtures of tetra-, penta-, and hexadecane, to adjust the melting temperature. Further data on mixtures of paraffins are listed in Kenisarin and Mahkamov 2007.

A *fatty acid* is characterized by the formula $CH_3(CH_2)_{2n}COOH$. In contrast to a paraffin, one end of the molecule ends with a –COOH instead of a –CH$_3$ group (fig.2.7).

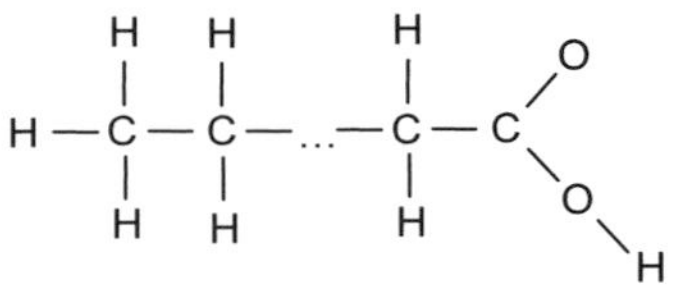

Fig. 2.7. Chemical structure of fatty acids.

Tab.2.6 lists some of the most common saturated fatty acids. Their melting enthalpy is similar to that of paraffins, and their melting temperature increases with the length of the molecule. Fatty acids are stable upon cycling; because they consist of only one component there cannot be phase separation. Like paraffins, fatty acids also show little or no subcooling and have a low thermal conductivity.

Table 2.6. Examples of fatty acids that have been investigated as PCM.

Material	Melting temperature (°C)	Melting enthalpy (kJ/kg)	Thermal conductivity (W/m·K)	Density (kg/m^3)
Caprylic acid $CH_3(CH_2)_6COOH$	16	149	0.149 (liquid, 38 °C) -	901 (liquid, 30 °C) 981 (solid, 13 °C)
Capric acid $CH_3(CH_2)_8COOH$	32	153	0.149 (liquid, 40 °C) -	886 (liquid, 40 °C), 1004 (solid, 24 °C)
Lauric acid $CH_3(CH_2)_{10}COOH$	42-44	178	0.147 (liquid, 50 °C) -	870 (liquid, 50 °C), 1007 (solid, 24 °C)
Myristic acid $CH_3(CH_2)_{12}COOH$	58	186, 204	- 0.17 (solid)	861 (liquid, 55 °C), 990 (solid, 24 °C)
Palmitic acid $CH_3(CH_2)_{14}COOH$	61, 64	185, 203	- -	850 (liquid, 65 °C 989 (solid, 24 °C)

A difference to paraffins can be expected in the compatibility of fatty acids to metals as experiments by Sari and Kaygusuz 2003 show; this is due to the acid character.

Different fatty acids can also be mixed to design PCM with different melting temperatures than the pure fatty acids. Data on such mixtures can be found in Nikolić et al. 2003, Tunçbilek et al. 2005, and Kenisarin and Mahkamov 2007.

Sugar alcohols are a hydrogenated form of a carbohydrate. The general chemical structure is $HOCH_2[CH(OH)]_nCH_2OH$, as fig.2.8 shows. Different forms are obtained depending on the orientation of the OH groups.

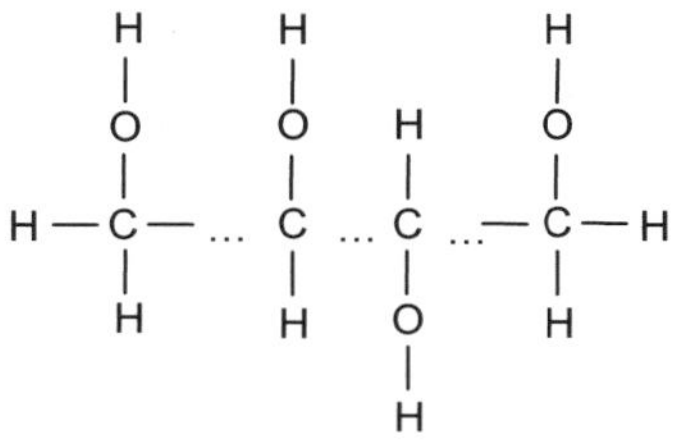

```
        H         H              H
        |         |              |
        O         O    H         O
        |         |    |         |
    H—C—  ...    C  ...C  ...  —C—H
        |         |    |         |
        H         H    O         H
                       |
                       H
```

Fig. 2.8. Chemical structure of sugar alcohols.

Sugar alcohols are a rather new material class, therefore little general information is available. As tab.2.7 shows, they have melting temperatures in the 90 °C to 200 °C range, and their mass specific melting enthalpies are comparatively high in most cases. In addition, their density is also high, which results in very high volume specific melting enthalpies. In contrast to many other organic materials, sugar alcohols however show some subcooling. According to Kakiuchi et al 1998, sugar alcohols are safe; erythritol and xylitol are used to replace sugar as sweetener.

Table 2.7. Examples of sugar alcohols that have been investigated as PCM.

Material	Melting temperature (°C)	Melting enthalpy (kJ/kg)	Thermal conductivity (W/m·K)	Density (kg/m^3)
Xylitol	94	263	-	-
$C_5H_7(OH)_5$			-	1500 (solid, 20 °C)
D-Sorbitol	97	185	-	-
$C_6H_8(OH)_6$			-	1520 (solid, 20 °C)
Erythritol	120	340	0.326 (liquid, 140 °C), 0.733 (solid, 20 °C)	1300 (liquid, 140 °C), 1480 (solid, 20 °C)
$C_4H_6(OH)_4$				
D-Mannitol	167	316	-	-
$C_6H_8(OH)_6$			-	1520 (solid, 20 °C)
Galactitol	188	351	-	-
$C_6H_8(OH)_6$			-	1520 (solid, 20 °C)

Polyethylen glycol, or short *PEG*, is a polymer with the general formula $C_{2n}H_{4n+2}O_{n+1}$. It is produced from ethylenglycol $C_2H_4(OH)_2$. The base unit of a linear PEG chain are monomers of $-CH_2-CH_2-O-$, as shown in fig.2.9.

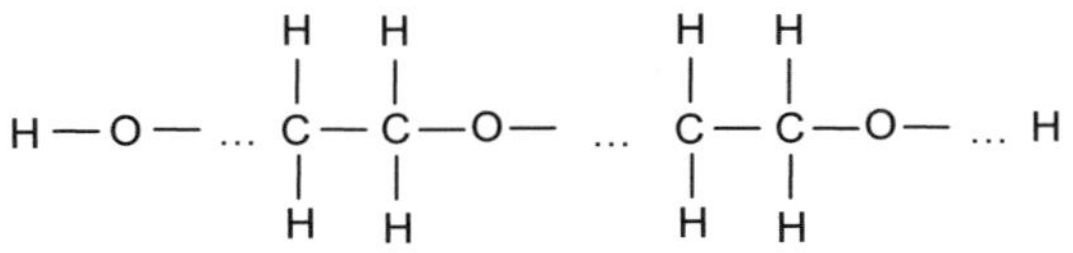

```
       H   H              H   H
       |   |              |   |
H—O— ...C — C—O—  ... C — C—O— ...H
       |   |              |   |
       H   H              H   H
```

Fig. 2.9. Chemical structure of polyethylene glycols.

The monomers have a molecular weight of 44 g/mole. Polyethylene glycols are available in a molecular weight range from about 200 to 35000; this corresponds to 5 monomers to about 800 monomers. Tab.2.8 lists some properties of PEGs.

Table 2.8. Examples of PEGs and some of their properties.

Material	Melting temperature (°C)	Melting enthalpy (kJ/kg)	Thermal conductivity (W/m·K)	Density (kg/m^3)
Diethylene glycol $C_4H_{10}O_3$	-10 to -7	-	-	1120 (liquid, 20 °C), -
Triethylene glycol $C_6H_{14}O_4$	-7	-	-	1120 (liquid, 20 °C), -
PEG400	8	100	0.19 (liquid, 38 °C), -	1125 (liquid, 25 °C), 1228 (solid, 3 °C)
PEG600	17 - 22	127	0.19 (liquid, 38 °C), -	1126 (liquid, 25 °C), 1232 (solid, 4 °C)
PEG1000	35 - 40	-	-	-
PEG3000	52 - 56	-	-	-
PEG6000	55 - 60, 66	190	-	1085 (liquid, 70 °C), 1212 (solid, 25 °C)
PEG10000	55 - 60	-	-	-

The various PEG types available have no precisely defined molar mass, but a certain molar mass distribution. PEGs with an average molecular weight between 200 and 400 are liquids at room temperature; PEG 600 melts at 17 – 22 °C. The melting temperature of all PEGs with a molecular weight exceeding 4000 g/mol is around 58 – 65 °C.

A class of materials that is not generally organic or inorganic is the clathrates. *Clathrates* are crystalline structures where molecules of one type are enclosed in the crystal lattice of another. Clathrates do not have a stochiometric composition; instead, there is an ideal composition when all free lattice positions are occupied. When the crystal lattice is formed by water, the clathrate is called a clathrate hydrate and the crystal structure of the solid clathrate hydrate is a modification of the crystal structure of ice. The molecules of the added substance are enclosed within free spaces in the crystal lattice of the ice and thereby stabilize it and raise the melting temperature. When the enclosed molecules are from a gas and the surrounding crystal lattice is water, the clathrate is called a gas hydrate. Gas hydrates have melting temperatures in the range from 0 °C to 30 °C, with the enclosed molecules being noble gases, chlorofluorocarbons, or straight chain hydrocarbons. The best-known gas hydrate is methane hydrate. Fig.2.10 shows its structure.

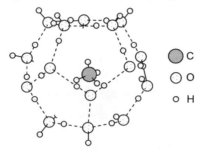

Fig. 2.10. Structure of methane hydrate: the methane molecule is located at the center of a cage of 20 water molecules

The gas molecules in a gas hydrate usually dissolve only at higher pressures in water and thus gas hydrates are often not stable at ambient pressure. This makes them difficult to use as PCM. Several exceptions however exist where the solubility in water is good and stability at ambient pressure is given. Other examples use special unpolar liquids or organic salts instead of gases. Tab.2.9 shows a few examples of clathrates.

Table 2.9. Examples of clathrates that have been investigated as PCM.

Material	Melting temperature (°C)	Melting enthalpy (kJ/kg)	Thermal conductivity (W/m·K)	Density (kg/m³)
tetrahydrofuran THF C_4H_8O + H_2O	5	280	- -	- 970 (solid)
tetrabutyl ammoniumbromide $[(CH_3(CH_2)_3]_4NBr$ + H_2O 40:60 wt.% (type A)	12	193	- -	- -

(Continued)

Table 2.9. (Continued)

Material	Melting temperature (°C)	Melting enthalpy (kJ/kg)	Thermal conductivity (W/m·K)	Density (kg/m^3)
tetrabutyl ammoniumbromide [(CH$_3$(CH$_2$)$_3$]$_4$NBr + H$_2$O 38:62 wt.% (type B)	10	199	-	-

Tetrabutylammonium bromide for example forms two different hydrates (Oyama et al. 2005), one with 26 water molecules per salt molecule (type A) and one with 38 (type B). Xiao et al. 2006 published further information on this hydrate. Like salt hydrates, clathrate hydrates may melt congruently or incongruently and tend to subcool. Further on, slow formation can be a problem.

As described until now, usually materials within the same material class are mixed. In the last few years, also mixtures of organic and inorganic materials have been investigated. However, at present not many results have been published. Therefore, this topic is not treated in this introductory text.

2.3 Typical material problems and possible solutions

Usually, a candidate material for a PCM does not fulfill all the requirements listed in section 2.1. Nevertheless, it is often still possible to use such a material if some of the strategies developed to solve or avoid potential problems are applied. Some of these strategies are now discussed.

2.3.1 Phase separation solved by mixing, gelling, or thickening

What is phase separation? When a pure substance with only one component, like water, is heated above its melting temperature and thereby melted, it will have the same homogeneous composition in the liquid as before in the solid (fig.2.11). When the material is solidified again by cooling it below the melting temperature, the solid will again be of the same homogeneous composition throughout and the same phase change enthalpy and melting temperature is observed at any place. Such a material is said to melt *congruently*.

2.3 Typical material problems and possible solutions 27

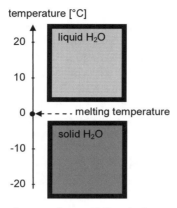

Fig. 2.11. Schematic phase diagram of water H₂O, a single component substance.

When a substance consists of two or more components, the phase diagram has to be extended. Fig.2.12 shows an example for a water-salt solution. For the second component, the salt, a new axis is added that gives the weight fraction of the components. In the case of two components, usually only one value is given; here it is the amount of salt in the salt solution. The second value, here for water, can be calculated as both fractions have to add up to 100 wt.%.

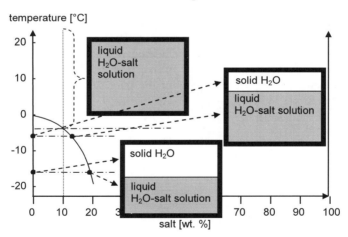

Fig. 2.12. Extension of the phase diagram when a second component, here a salt, is added to water.

The two-component system now behaves very different. A salt-water solution with a composition of 10 wt.% salt and 90 wt.% of water, marked by the vertical dotted line in fig.2.12, is a homogeneous liquid above -4 °C. When cooled below -4 °C, water freezes out of the solution and consequently the remaining solution has higher salt concentration. This means the substance separates into two different phases, one with only water, and a second one with a higher salt concentration than initially. Due to gravitation, the phase with higher density will sink to the bottom and the one with lower density to the top. This phenomenon is called *phase separation* or *decomposition*, because the original composition is changed. When the temperature is reduced further, more water freezes out, and the salt concentration in the remaining liquid increases, as shown in fig.2.12. For different initial compositions, the temperature where water starts to freeze out is also different: the higher the salt concentration, the lower the temperature where water starts to freeze out of the solution.

The line that separates the parts of the phase diagrams where there is only a homogeneous liquid from others is called liquidus line. Fig.2.12 only shows a small fraction of the liquidus line for compositions with low salt concentration. For high salt concentrations, the liquidus line must rise with increasing salt concentration again. The reason is that the melting temperature of a salt is higher than for water and with the addition of small amounts of water to the salt the salt will melt at a lower temperature. Pure water and pure salt have locally high melting temperatures and high phase change enthalpies. If salt is added in a small amount to water, or if water is added in a small amount to salt, the liquidus line moves to lower temperatures and the enthalpy difference between two temperatures is also reduced.

Another energetically stable mixture are salt hydrates, which consist of a salt and water in a discrete mixing ratio salt·nH_2O. Fig.2.13 shows the phase diagram for a salt hydrate with congruent melting. The composition in the example is 50 wt.% salt and 50 wt.% water. If the homogeneous liquid with this composition is cooled, a single solid phase is formed from the liquid: the salt hydrate. This is the same as for pure water or for the pure salt, and again, there is no phase separation.

2.3 Typical material problems and possible solutions 29

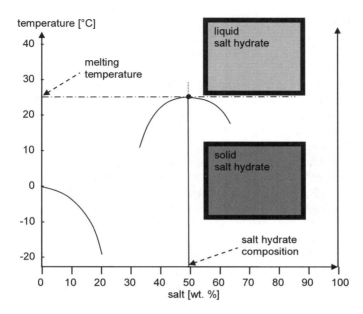

Fig. 2.13. Phase diagram for a salt hydrate, which shows congruent melting.

The liquidus lines from the water with added salt, and from the salt hydrate with less water, have to meet somewhere. This could be at an eutectic point; the respective composition is called eutectic composition. Fig.2.14 shows this case.

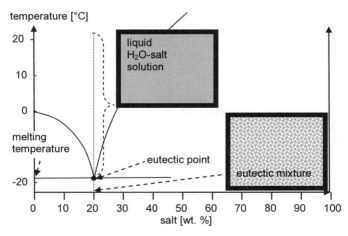

Fig. 2.14. Phase diagram for an eutectic composition, which shows congruent melting.

Eutectic mixtures are mixtures of two or more constituents, which solidify simultaneously out of the liquid at a minimum freezing point, also called *eutectic point*. At the eutectic point, the liquid reacts to a solid that is composed of two or more solid phases with different composition; however, the overall composition is still the same as in the liquid! Therefore, eutectic compositions do not show phase separation. This is the reason why the salt-water solutions used as PCM (fig.2.2, tab.2.1) all have an eutectic composition. Strictly speaking, eutectic compositions are not under all circumstances congruent melting. If there is subcooling, the different solid phases will not be formed simultaneously out of the liquid, and phase separation can occur. However, this is usually not the case (Lane 1983).

After discussing the most important cases for congruent melting, it is time to discuss those where melting is not congruent. The most commonly used salt hydrates cannot be described by the phase diagram shown in fig.2.13. Often, compositions of salt·nH_2O with different n exist, and the associated liquidus lines overlap. This can lead to a complex phase diagram and different melting behavior. An example is $CaCl_2·6H_2O$, which is said to melt semi congruent. According to the schematic phase diagram shown in fig.2.15, $CaCl_2·6H_2O$ has a composition of about 50 wt.% of $CaCl_2$ and 50 wt.% of water. The cool down of a liquid of this composition is indicated by arrows. When the liquid is cooled, solid $CaCl_2·6H_2O$ does not form directly. Instead, a solid and a liquid are formed first. The solid has the composition of $CaCl_2·4H_2O$ and the liquid consequently has a higher water content than $CaCl_2·6H_2O$. The reason for this behavior is that the liquidus curve of $CaCl_2·4H_2O$ intersects the one of $CaCl_2·6H_2O$ at a lower salt concentration.

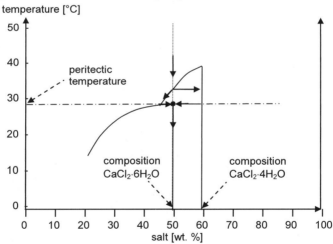

Fig. 2.15. Schematic phase diagram of $CaCl_2$-H_2O, which shows semi congruent melting.

When a temperature of about 28 °C is reached, the solid $CaCl_2·4H_2O$ and the liquid salt solution react at a fixed temperature to a single solid phase: $CaCl_2·6H_2O$. This transformation of a solid and a liquid to a solid is called a *peritectic transformation*, in contrast to eutectic transformations where a liquid transforms to two or more solid phases. The corresponding temperature is called *peritectic temperature*. When solid $CaCl_2 \cdot 6H_2O$ is heated again above the peritectic temperature of about 28 °C, first solid $CaCl_2 \cdot 4H_2O$ and liquid salt solution is formed. When a temperature a few Kelvin above the peritectic temperature is reached, a homogeneous liquid can form again. The problem of peritectic transformations is that in the intermediate step phases with different densities show up. Here, these phases are solid $CaCl_2 \cdot 4H_2O$ with less water, which sinks down, and a solution with less $CaCl_2$. Only when the temperature is above the liquidus curve long enough, solid $CaCl_2 \cdot 4H_2O$ can dissolve and a homogeneous liquid is formed. If cooling is done slowly, the different phases can build up and separate again. If cooling however is fast enough, or if macroscopic separation of the two phases is avoided by some other means, solid $CaCl_2 \cdot 6H_2O$ can be formed again from the liquid and the phase change enthalpy is released completely. Such a melting behavior that is not congruent and not incongruent melting is called *semi congruent melting*.

The problem with phase separation is that it can severely reduce the storage density. Because locally the right concentration of the molecules to form the PCM is not given anymore, the PCM cannot solidify throughout. This means the latent heat of solidification can usually not be released completely, sometimes only to a small fraction. To retrieve the latent heat stored in the initial material, the correct concentration of the chemical components is required throughout the whole sample. Only then, the PCM can solidify completely.

An example where phase separation is worse than in $CaCl_2 \cdot 6H_2O$ is the salt hydrate $Na_2SO_4 \cdot 10H_2O$, also called Glaubersalt. Glaubersalt is composed of Na_2SO_4 and 10 molecules of H_2O, that is equal to about 43 wt.% of Na_2SO_4 (fig.2.16). When solid $Na_2SO_4 \cdot 10H_2O$ is heated up, the composition follows the vertical line in the phase diagram at 43 wt.%. When the peritectic point of 32 °C is reached, again two phases are formed: Na_2SO_4 and a liquid, which contains only about 32 wt.% Na_2SO_4. The two phases have very different densities and the pure Na_2SO_4, the one with the higher density, sinks down. Up to this point, the behavior is similar to the one of $CaCl_2 \cdot 6H_2O$. When the temperature is raised further, the concentration of Na_2SO_4 in the liquid however decreases further because in contrast to the case of $CaCl_2 \cdot 6H_2O$ the solubility of Na_2SO_4 decreases with rising temperature. This means it is not possible to get a homogeneous liquid with the original composition by raising the temperature. When the temperature is lowered again below 32 °C, $Na_2SO_4 \cdot 10H_2O$ will form again from the already diluted solution and sink down with an even more diluted solution remaining.

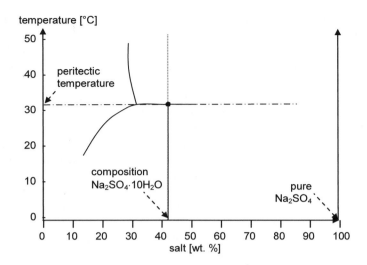

Fig. 2.16. Schematic phase diagram of Na_2SO_4-H_2O, which shows incongruent melting.

From equilibrium thermodynamics, $Na_2SO_4 \cdot 10H_2O$ is still the most stable phase below the melting temperature. However, to dissolve Na_2SO_4 by molecular diffusion in the solution will take a very long time because it is separated from the solution by a layer of $Na_2SO_4 \cdot 10H_2O$.

Depending on how severe phase separation occurs and how difficult it is to get rid of it, the melting behavior of a PCM showing phase separation is described as *semi congruent melting* or as *incongruent melting*, in contrast to congruent melting if no phase separation occurs. The use of these words however does not follow an exact definition and therefore only roughly indicates how a material will perform. In this discussion, it is also important to keep in mind that phase diagrams are supposed to show data in thermodynamic equilibrium. This means two things: first that the temperature is the same throughout the bulk of the sample, and second, that dynamic processes like molecular diffusion and reactions have come to a halt. This also includes the effect of subcooling. Because data in phase diagrams are taken from measurements, this ideal case can only be approached (section 3.2.2). Therefore, phase diagrams can be helpful to understand how a sample will perform, but there is no guarantee that they show all effects. Depending on the heating and cooling rate, the degree of phase separation in semi congruent and incongruent melting PCM will be different. This explains that often for the same material different literature sources give different melting behavior.

The question is now: if a candidate material for a PCM has a potential for phase separation, how is it possible to get rid of that problem? If phase separation has occurred, of course *artificial mixing* can be used. This is a well-known approach to dissolve sugar in coffee, or salt in water. Applied to PCM, the PCM is allowed to separate on macroscopic distances, but instead of waiting for diffusion to homogenize the PCM, the faster process of mixing is used. This approach has been used successfully with many salt hydrates. Its main disadvantage is the necessary equipment. Because artificial mixing is not a solution on the material level but on the level of the storage concept, this approach is discussed in chapter 5.

An easy approach on the level of the material is to add additional water to the salt hydrate. For example instead of $CaCl_2 \cdot 6H_2O$ with about 50 wt.% of $CaCl_2$ a more diluted mixture with less than 45 wt.% can be used. The phase diagram in fig.2.15 shows, that upon heating the formation of $CaCl_2 \cdot 4H_2O$ is now avoided, because the intersection of the two liquidus lines is passed on the left side. Further on, the additional water will make the homogenization of the liquid phase by diffusion of $CaCl_2 \cdot 4H_2O$ faster. The drawbacks of this method are that due to the addition of water, the overall storage density is reduced, and that the melting range becomes broader.

A second way to reduce the problem of phase separation on the material level is by using diffusion processes for homogenization. Diffusion is however only efficient on small scales, because the speed of diffusion processes goes with the square of the distance. Therefore, this approach can work only if the PCM separates only on small distances. One way to limit the distance that the phases can separate to the scale of several mm is to use shallow containers for the PCM. But often this is not sufficient. To reduce the distance that the phases can separate down to a microscopic scale, *gelling* can be used. In gelling, a three dimensional network is formed within the bulk of the PCM. This network holds the different phases of the PCM together on a microscopic scale. The gel can be formed by a polymer, for example. Fig.2.17 shows water gelled with a cellulose derivative. The same effect as with gelling can be achieved if the PCM is infiltrated into a micro porous material.

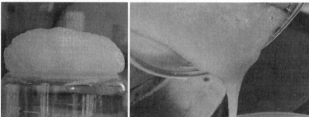

Fig. 2.17. Water gelled with a cellulose derivative, left, and thickened salt hydrate, right (pictures: ZAE Bayern).

Another way to reduce the distance that the phases can separate is by *thickening* the PCM. Thickening means the addition of a material to the PCM to increases its viscosity. Due to the high viscosity, different phases cannot separate far until finally the whole PCM is solid. Fig.2.17 shows a thickened salt hydrate. There is a third way to reduce phase separation on the material level; it is probably the best but also most complicated one. This way is changing the phase diagram of the PCM itself by the addition of other materials until congruent melting results. If achieved, the PCM can be used without any restrictions. However if and how the melting behavior can be changed advantageously is a topic for experts in the field.

For readers interested in more details, the following literature can be recommended. A very good introduction into phase diagrams can be found in Atkins et al. 2002. Phase equilibriums, phase diagrams and how they are determined, and the dynamics of the formation of different phases, is described in Lane 1983. The discussion also includes an in depth description of supersaturation and subcooling, nucleation and crystallization, and nucleators. Thickening and gelling are described in Lane 1986. Farid et al. 2004 give examples of thickeners for several important PCM.

2.3.2 Subcooling and methods to reduce it

Many PCM do not solidify immediately upon cooling below the melting temperature, but start crystallization only after a temperature well below the melting temperature is reached. This effect is called *subcooling* or supercooling (fig.2.1). For example, liquid water can be cooled to temperatures well below 0 °C; if highly pure and in small quantities even below -15 °C. Fig.2.18 shows the effect of subcooling on heat storage.

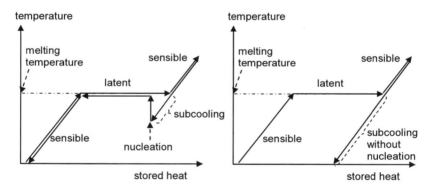

Fig. 2.18. Effect of subcooling on heat storage. Left: with little subcooling and nucleation, right: severe subcooling without nucleation.

During the supply of heat, there is no difference whether a PCM shows subcooling or not. During extraction of heat however, the latent heat is not released when the melting temperature is reached due to subcooling. The effect of subcooling makes it necessary to reduce the temperature well below the phase change temperature to start crystallization and to release the latent heat stored in the material. If nucleation does not happen at all, the latent heat is not released at all and the material only stores sensible heat. In technical applications of PCM, subcooling therefore can be a serious problem.

When water is subcooled to -8 °C and crystallization starts, the latent heat of crystallization is released. At that temperature of -8 °C, 8 K · 4 kJ/kg·K = 32 kJ/kg of sensible heat were lost due to subcooling, much less than the latent heat of 333 kJ/kg which is released during crystallization. If the heat released upon solidification is larger than the sensible heat lost due to subcooling, the temperature rises to the melting temperature and stays there until the phase change is completed. This case is shown at the right in fig.2.18. If the loss of sensible heat during subcooling is however larger than the latent heat released upon crystallization, or if the rate of heat loss to the ambient is larger than the rate of heat release during crystallization, it is possible that the temperature will not rise to the melting temperature again. This means that subcooling can cause surprising effects when performing a dynamic experiment. Fig.2.19 shows the effect of subcooling on a cool down experiment with NaOAc·3H$_2$O. After the crystallization has started at a temperature of about 55 °C, the temperature rises sharply until the melting temperature of about 57 °C is reached. When the melting temperature is reached at a location, no more phase change enthalpy is released and the temperature rise is stopped. From the location where the first crystallization occurs, the phase front moves until the boundaries of the sample are reached and phase change takes place throughout. This usually leads to a longer or shorter plateau in the temperature at the melting temperature. The length of this plateau depends on the rate of heat loss at the boundaries of the sample to the environment. In fig.2.19, the plateau is about 1 hour long before the temperature goes down to the temperature of the environment, which is 40 °C.

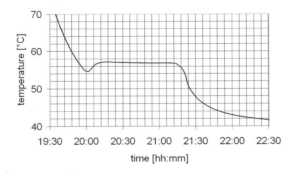

Fig. 2.19. Cooling of NaOAc·3H$_2$O at an ambient temperature of 40 °C with about 2 K subcooling (picture: ZAE Bayern).

What is the reason for subcooling, or better why does a material not solidify right away when cooled below the melting temperature? Solidification means that the amount of solid phase grows because the liquid phase at the interface with the solid phase solidifies. However, this has to start at some point and the starting can be a problem. At the very beginning, there is no or only a small solid particle, also called nucleus. For the nucleus to grow by solidifying liquid phase on its surface, the system has to release heat to get to its energetic minimum. There is however a competition between the surface energy, which is proportional to the surface area and therefore to r^2 (r is the radius of the nucleus), and the heat released by changing the phase, which is proportional to the volume and therefore to r^3. At small radius, it is possible that the heat released by crystallization is smaller than the surface energy gained. That means there is an energetic barrier; only when a nucleus of sufficiently large radius is present solidification starts. For this, in some cases temperatures significantly lower than the melting temperature are necessary. Based on this, nucleation is divided into two cases:

- *Homogeneous nucleation* means nucleation solely started by the PCM itself.

  Homogeneous nucleation includes nucleation by low enough temperatures and a second possibility, that particles of the solid PCM are added to the subcooled PCM. The latter one is sometimes also called *secondary nucleation*.

- *Heterogeneous nucleation* means nucleation not by the PCM itself.

  Origins of heterogeneous nucleation can be special additives intentionally added to the PCM, but also impurities, or cracks at the wall of the vessel that contain solid PCM.

The most common approach to get rid of subcooling on the level of the PCM is to add special additives, also called *nucleator*, to the PCM to cause heterogeneous nucleation. Nucleators have been developed for most well investigated PCM, and reduce subcooling typically to a few K (fig.2.19). Most nucleators are materials with a similar crystal structure as the solid PCM to allow the solid phase of the PCM to grow on their surface, but a higher melting temperature to avoid deactivation when the PCM is melted. The problem with this method is that usually a similar crystal structure also means a similar melting temperature. Therefore, many nucleators are only stable up to a temperature 10 K to 20 K above the melting temperature of the PCM. There are also other nucleators where the mechanism is completely unknown. The fact that there is still no reliable theoretical approach makes the search for a new nucleator time consuming. Examples of nucleators for some of the most important PCM are given in Farid et al. 2004 and in Lane 1983. Lane 1983 also gives an in depth description of nucleation, crystallization, and nucleators. Other sources of general information on nucleation, not specialized on PCM, are the books by Kashchiev 2000, Kurz and Fisher 1992, Mutaftschiev 1993, Pimpinelli and Villain 1998, and Herlach 2004.

The difficulty of finding a nucleator can be circumvented if the PCM is locally cooled to a sufficiently low temperature to always have some solid phase left that acts as homogeneous nucleator. The necessary local cooling can be done by a Peltier element or by the *cold finger* technique. The cold finger technique uses a cold spot in the containment, caused for example by an intentionally bad insulation, to always have some solid PCM. It is therefore only useful when the melting temperature of the PCM is higher than the temperature of the ambient.

Until recently, it was a common theory that the metal clip used in pocket heaters causes nucleation by a shock wave. The shock wave emitted when the metal clip is bent was thought to locally cause a high pressure in the PCM and thereby start nucleation. Rogerson and Cardoso 2003a-c however proved that this theory was wrong. They could show that surface effects stabilize solid PCM particles in small cracks of the metal clip in a way that the melting temperature of these PCM particles is much higher than the melting temperature of the bulk PCM. When the metal clip is bent, these solid PCM particles are ejected into the bulk of the PCM and initiate homogeneous nucleation. Nevertheless the option to start solidification using the high pressure associated with shock waves is very attractive and currently investigated in the project LWSNet supported by the German Federal Ministry of Education and Research (BMBF) under the funding code 03SF0307A (http://www.lwsnet.info/objective.htm). First experimental results published by Günther et al. 2007 have shown the general feasibility to nucleate salt hydrates by high pressures. In these experiments, static pressures of several kbar were used. The necessary equipment to produce sufficiently high static pressures is however prohibitive in PCM applications. If dynamic pressures in pressure waves, which can be produced with less effort, also cause nucleation will be investigated in future experiments.

2.3.3 Encapsulation to prevent leakage and improve heat transfer

In most cases, except for some applications of water-ice, the PCM needs to be encapsulated. The two main reasons are to hold the liquid phase of the PCM, and to avoid contact of the PCM with the environment, which might harm the environment or change the composition of the PCM. Further on, the surface of the encapsulation acts as heat transfer surface. In some cases, the encapsulation also serves as a construction element, which means it adds mechanical stability. Encapsulations are usually classified by their size into macro- and microencapsulation.

Macroencapsulation means filling the PCM in a macroscopic containment that fit amounts from several ml up to several liters. These are often containers and bags made of metal or plastic. Macroencapsulation is very common because such containers or bags are available in a large variety already from other applications. In that case, macroencapsulation is mainly done to hold the liquid PCM and to prevent changes in its composition due to contact with the environment. If the container is rigid enough, the encapsulation can also add mechanical stability to a system.

Microencapsulation is the encapsulation of solid or liquid particles of 1 µm to 1000 µm diameter with a solid shell. Physical processes used in microencapsulation are spray drying, centrifugal and fluidized bed processes, or coating processes e.g. in rolling cylinders. Chemical processes used are in-situ encapsulations like complex coacervation with gelatine, interfacial polycondensation to get a polyamide or polyurethane shell, precipitation due to polycondensation of amino resins, and others. The in-situ processes have the ability to yield microcapsules with the best quality in terms of diffusion-tightness of the wall. Such microcapsules are nowadays widely used in carbonless copy paper (Jahns 1999). Besides the containment of the liquid phase, other advantages of microencapsulation regarding PCM are the improvement of heat transfer to the surrounding because of the large surface to volume ratio of the capsules, and the improvement in cycling stability since phase separation is restricted to microscopic distances. Further on, it is also possible to integrate microencapsulated PCM into other materials. A potential drawback of microencapsulation is however that the chance of subcooling increases. Currently, microencapsulation on a commercial scale is applied only to PCM that are not soluble in water. The reason is the used process technology. Besides that, to microencapsulate salt hydrates there is an additional problem: the tightness of the shell material to the small water molecules has to be sufficient to prevent changes of the composition of the salt hydrate. Two chemical processes for micro encapsulation are now described in more detail.

In *simple and complex coacervation* many products can be used as wall materials. For example, urea-formaldehyde resin, melamine-formaldehyde resin, β-naphtol-formaldehyde resin and gelatine-gum Arabic (this last one for complex coacervation) have been used. The first step of microencapsulation by coacervation is to disperse the core material (PCM) in an aqueous gelatine solution at a temperature range of 40 – 60 °C, at which the solution of the wall material is liquid. In complex coacervation, gum Arabic is added to the gelatine at this point. The process is followed by adjusting the pH and concentration of polymer so that a liquid complex coacervate is formed (usually pH 4 to 4.5). When the liquid coacervate is formed, the system is cooled to room temperature. The final step is hardening and isolation of the microcapsules. Hardening can be done using formaldehyde, which crosslinks the wall material by reacting with amino groups located on the chain. After that, the pH should be raised to 9 - 11 by NaOH solution. Then the capsules are cooled approximately down to 5 – 10 °C and maintained at this temperature for 2 to 4 h. Then the capsules are filtered and dried (Özonur et al. 2006, Jahns 1999).

Polymerization is a technique usually performed using urea and formaldehyde, in an oil-in-water emulsion. First, at about 20 °C to 24 °C, deionized water and an aqueous solution of ethylene maleic anhydride (EMA) copolymer are mixed in a vessel with a controlled temperature. While mixing, urea, ammonium chloride, and resorcinol are dissolved in the solution. The pH is raised from about 2.60 to 3.50 by drop-wise addition of sodium hydroxide (NaOH) and hydrochloric acid (HCl). A slow stream of PCM is then added to form an emulsion and allow stabilizing for 10 min. After stabilization, an aqueous solution of formaldehyde is added. The emulsion is heated to the target temperature of 55 °C and stays there for 4 h under continuous mixing. Once cooled to ambient temperature, the suspension of microcapsules is separated under vacuum with a coarse-fritted filter. The microcapsules are rinsed with deionized water and air dried for 24 – 48 h. A sieve can be used to separate the microcapsules (Brown et al 2003).

A good and easy to understand overview on techniques used for microencapsulation is available at the website of Microtek Laboratories at http://www.microteklabs.com.

2.3.4 Mechanical stability and thermal conductivity improved by composite materials

A PCM can be combined with other materials to form a composite material with additional or modified properties. A composite can be formed in different ways, shown in fig.2.20: by embedding another material into the PCM or by embedding the PCM into a matrix of another material.

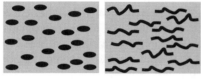 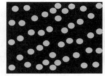 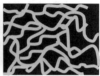

        particles and fibers   PCM (grey) embedded in a matrix ma-
    embedded in a PCM (grey)   terial with pores or channels
examples: examples:
- graphite fibers   - graphite matrix
- metallic particles and fibers   - metallic matrix / foam
    - polymer foam

Fig. 2.20. Possibilities to form a composite material, and some examples.

To maintain material properties, the order of magnitude of the structures in the composite should be microscopic, or at least below the scale of mm. Otherwise, the properties of the composite will depend on the sample size and the composite therefore cannot be called a material anymore (section 3.1).

2.3.4.1 Mechanical stability

Composites of a PCM and another material as a mechanically stable structure are often called shape-stabilized PCM (ss-PCM). Independent of the phase of the PCM, solid or liquid, the shape is maintained by the supporting structure. An example is shape-stabilized paraffin, which can be produced by incorporating paraffin on a microscopic level into a supporting structure, for example high-density polyethylene (HDPE). Inaba and Tu 1997 describe the preparation of samples with about 75 wt.% of paraffin and report that there is no leakage of the paraffin out of the HDPE structure. Recently, Yinping et al. 2006 have published data with even 80 wt.% and other physical properties of the composite. They also report that no containment is needed. Compared to the encapsulation techniques described before, these new materials could be cut into arbitrary shapes without leakage. This is a significant advantage in many applications, however at the cost of using much more HDPE than if just as a wall material for a macroencapsulation. Farid et al. 2004 also report on the composite between HDPE and paraffin in their review paper. Further on, they mention a development published by Royon et al 1997, who have developed a new material where water as a PCM is integrated into a three dimensional network of polyacrylamide during the polymerization process.

Another approach to improve the mechanical stability by forming a composite material is the impregnation of mechanically stable, porous materials with the PCM. For example, ceramic granules and tiles have been infiltrated with salts as PCM to get storage modules for temperatures of several hundred °C. Another example is the impregnation of wood fiberboards with paraffin.

It is also possible to combine both approaches. The company Rubitherm Technologies GmbH developed several materials that consist of three components: a mechanically very stable and highly porous structure, whose pores are filled with a paraffin as PCM, and a polymer structure to keep the paraffin within the pores. As mechanically stable and highly porous structure ceramic or silica powders and wood fiberboards are used.

2.3.4.2 Thermal conductivity

All non-metallic liquids, including PCM have a low thermal conductivity (tab.2.1 to tab.2.8). Since PCM store large amounts of heat or cold in a small volume, and because it is necessary to transfer this heat to the outside of the storage to use it, the low thermal conductivity can be a problem. In the liquid phase, convection can significantly enhance heat transfer, however often this is not sufficient. In the solid phase, there is no convection. When fast heat transfer is necessary, one possibility to increase the thermal conductivity of the PCM is to add materials with larger thermal conductivity. This can be done at a macroscopic scale, for example by adding metallic pieces (Velraj et al. 1999, Cabeza et al. 2002, and Hafner and Schwarzer 1999), or on a sub mm scale as mentioned before with composite materials. However, adding anything to the PCM will reduce or eliminate convection in the liquid phase; therefore, it is necessary to find out which option is better. One approach under investigation to increase the thermal conductivity is to put the PCM into metallic foams. In recent years, the technology to produce metal foams of different porosity and structure has advanced considerably. Hackeschmidt et al. 2007 have prepared and tested combinations of water and a paraffin as PCM with aluminum foams of different porosity. The results show that with a relative density of 6 %, that means with 94 % porosity, a thermal conductivity of about 6 W/mK can be achieved. Other experiments are described by Hong and Herling 2006. The use of graphite and its potential to increase the thermal conductivity when paraffins or salt hydrates are used as PCM is described in Mehling et al. 1999, Mehling et al. 2000, Py et al. 2001, and Mills et al. 2006. Recently, the approach to use graphite is also investigated for salts as high temperature PCM (Bauer et al 2006, do Couto Aktay et al. 2005). For PCM-graphite composites with up to 85 vol.% of PCM thermal conductivities of 2 to 30 W/m^2K have been reported.

Besides the approach on the level of the storage material, the problem of heat transfer can also be approached on the level of the storage concept. This is discussed in chapter 5.

2.4 Commercial PCM, PCM composite materials, and encapsulated PCM

The availability of commercial PCM, PCM composite materials and encapsulated PCM is crucial to the development and commercialization of PCM applications. The reason is that from a customers point of view, only commercial PCM, PCM composite materials and encapsulated PCM have defined properties, a warranty, a fixed price, and can be delivered in a given time. From a supplier point of view, the size of the potential market of a PCM, PCM composite, or encapsulated PCM is also important. Both views determine what is commercially available. Commercial PCM and PCM composite materials have to fulfill harder requirements in their development than encapsulated materials (Mehling 2001). Often a potential material cannot be sold as pure PCM because handling of the pure material is critical, for example with respect to its water content. In an encapsulated form, such materials nevertheless can be sold as product. Looking at large systems, the situation can be the opposite: a PCM that shows phase separation might not be useful as encapsulated PCM, however when artificially mixed in a large tank it can be used. Then however, the applicability only to large systems strongly restricts the market for such a PCM unless the system uses large amounts of the PCM or the application is widespread.

The availability of commercial PCM, PCM composite materials, and encapsulated PCM allows companies with no or little knowledge on PCM to use this technology in their own products. Encapsulated PCM can be commercialized by one company specialized on PCM technology, and many other companies can then integrate the encapsulated PCM in their products, without knowing much about PCM technology. A good example is ice packs, which are used by many consumers to keep food cold.

Usually, it is necessary that the main properties of the PCM products are well documented. For this reason, a standard to control product quality has been developed by the ZAE Bayern and the FhG-ISE recently. Since spring 2007, the quality label shown in fig.2.21 will indicate that a PCM product has been tested according this standard.

Fig. 2.21. Quality label for PCM (picture: Gütegemeinschaft PCM e.V.)

More information on the standard and the label can be found at http://www.pcm-ral.de/. Similar activities to develop standards have recently been started in China and in India.

The number of commercial PCM, PCM composite materials, and encapsulated PCM is growing from year to year. Therefore, it is not possible to give a complete description of all available commercial products here. The following sections discuss typical examples to give an overview on what is available; for many examples similar products are available from several companies. Detailed and up to date information is available at the respective websites of the companies listed in the reference section at the end of this chapter.

2.4.1 PCM

Currently, more than 50 PCM are commercially available. Fig.2.22 shows an overview of the phase change temperature and enthalpy per volume and mass of these commercial PCM.

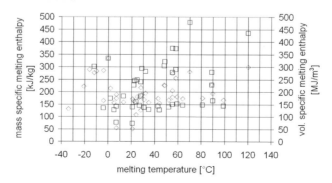

Fig. 2.22. Phase change temperature and enthalpy per volume (□) and mass (◊) of commercial PCM (picture: ZAE Bayern)

Most commercial PCM are based on materials from the material classes of the salt hydrates, paraffins, and eutectic water-salt solutions. They are however not identical with these materials. In the case of salt hydrates, often the composition is changed, a nucleator is added, the material is gelled or thickened, or the PCM is a mixture of different base materials. With paraffins, commercial grade paraffins usually contain a mixture of different alcanes because pure alcanes are expensive. Commercial PCM cover the temperature range from − 40 °C to +120 °C. Even though many materials have been investigated for higher temperatures, none of them is available commercially because there has been no market yet.

The price of commercial PCM is typically in the range from 0.5 €/kg to 10 €/kg, which has a large influence on the economics of PCM applications. For a rough estimate an energy price of 0.05 €/kWh for heat can be assumed. This means that 3600 kJ cost 0.05 €. Taking an average storage density of a PCM of 180 kJ/kg (fig.2.22), 20 kg of PCM are necessary to store 3600 kJ (=1 kWh), an amount of heat that has a value of 0.05 €. 20 kg of PCM however cost at least 20 kg · 0.5 €/kg = 10 €. To store heat with a value that equalizes the cost of the necessary investment for the PCM, a number of 10 € / 0.05 € = 200 storage cycles are necessary. Additional investment cost for the storage container and heat exchanger, as well as the stored heat which is also never completely free, have not even been taken into account. Seasonal storage using PCM is therefore far from being economic at current prices for fossil fuels. To be competitive in energy systems, one should try to charge and discharge a storage daily, or in even shorter periods. There are however exceptions. For example if an application has no connection to the energy grid, e.g. in the cargo bay of an airplane, the common energy price is not applicable and the economic situation can be much better. The examples in chapter 7 and 8 reflect this situation.

2.4.2 PCM composite materials

A composite material is a material that is composed of several different materials, usually to improve a property of a material or to combine properties of different materials. In the case of PCM, a PCM composite material is produced to improve at least one of the PCM properties or to improve the heat storage capacity of another material. There are different ways to form a composite: by embedding PCM in a matrix of another material, or by embedding another material into the PCM (section 2.3.4). The following section gives some examples of composite materials developed to improve different properties of PCM.

2.4.2.1 PCM composite materials to improve handling and applicability

A set of different composite materials with paraffin as PCM, mainly to improve handling and integration into a product, has been developed and commercialized by the company Rubitherm Technologies GmbH. The compound PK (fig.2.23) is a shape stabilized PCM and consists of a paraffin as PCM in a polymer structure.

Fig. 2.23. Composites developed and commercialized by Rubitherm Technologies GmbH. From left to right: compound PK, powder PX, and granulate GR (pictures: Rubitherm Technologies GmbH).

The powder PX and granulate GR have an additional mechanically stable structure. The powder RUBITHERM® PX is based on a silica powder as matrix material, which contains up to 60 % PCM and still flows as the pure powder. It is used to fill complicated or flexible shapes, for example in comfort and medical applications as the hot cushion shown in fig.8.9. The granulate RUBITHERM® GR is based on a porous clay mineral and contains about 35 % of PCM. The granulate can be used to fill complex shapes, like the powder, but it is also mechanically very stable. It can even be used in a floor heating system (fig.9.54).

Fig.2.24 shows examples of composites that come as boards or plates. The fiberboard RUBITHERM® FB contains about 65 % of PCM and is used for example for plates to keep Pizza or other food warm. Here the shape as a plate is important.

Fig. 2.24. Composites that come as boards or plates. Left: fiberboard FB (picture: Rubitherm Technologies GmbH) and right: DuPont™ Energain® panels (picture: DuPont)

DuPont recently announced the commercialization of a composite to be used as building material. DuPont™ Energain® panels contain a copolymer and paraffin compound, with about 60 wt.% of paraffin. Section 9.2.1.4 contains more data on the composite and its application.

2.4.2.2 PCM-graphite composites to increase the thermal conductivity

If the thermal conductivity of a material is not large enough, it is a common strategy to form a composite with a highly conducting material. In many applications, metal or graphite is used as additive in the form of powders or fibers. Besides the high thermal conductivity of graphite, its stability to high temperatures and corrosive environments is a big advantage compared to metals. This is especially the case when applied in combination with salts, salt hydrates, water, or water-salt solutions. The idea to use a graphite matrix as thermally conducting structure was developed and patented by the ZAE Bayern, and first published by Satzger et al. 1997. Since then, a number of publications (Mehling et al. 1999, Mehling et al. 2000, Py et al. 2001, Mills et al. 2006) have presented the approach as well as experimental results. PCM-graphite composites in two different forms, a matrix and a compound material, are available commercially nowadays under the brand name ECOPHIT™ from SGL TECHNOLOGIES GmbH. A short information sheet is available at http://www.sglcarbon.com/sgl_t/expanded/pdf/Ecophit_LC_GC_d.pdf.

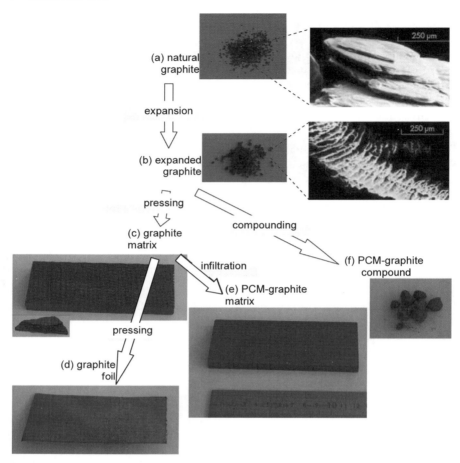

Fig. 2.25. Processing of graphite from natural graphite to different products. All pictures have the same scale (microscopic images of expanded graphite: SGL TECHNOLOGIES GmbH).

Fig.2.25 shows the processing from natural graphite to different products. The first processing step starts with natural graphite (a) with a density of about 2200 kg/m^3. After a chemical treatment, the following thermal treatment of the natural graphite starts a chemical reaction, which produces gases. These gases lead to an expansion of the graphite structure as the microscopic image in fig.2.25 shows. The particles of the expanded graphite (b) have a highly porous structure and dimensions in the order of one to several mm. This gives them a bulk density of about 3 kg/m^3, less than 0.2% of the density of natural graphite. In other words, 99.8% of the volume is pores.

PCM-graphite matrix

In a standard production process, graphite seals for applications at high temperatures or in corrosive environments are produced. For this, the expanded graphite is compressed in several steps to graphite foil (d). During compression, the density increases by a factor of about 700 from about 3 kg/m^3 for the expanded graphite, to a density close to the one of pure graphite of 2.2 kg/m^3. This is why these foils can be used as seals. In an intermediate step of the compression, the graphite has the shape of about 10 mm thick plates (c). These plates still have about 90 vol.% porosity, but yet good mechanical stability and a thermal conductivity of about 20 to 25 W/mK parallel and 5 to 8 W/mK perpendicular to the plate surface. Compared to this, organic PCM typically have a thermal conductivity of 0.2 W/mK or less; inorganic PCM typically have 1.0 W/mK or less. Because of their high porosity and mechanical stability, the prepressed plates are called a graphite matrix. In one process to produce PCM-graphite composite materials, this matrix is used as thermally conductive matrix. The PCM is infiltrated into the graphite matrix until about 80 to 85 vol.% of PCM are reached. Fig.2.25 shows the PCM-graphite matrix (e). Fig.2.26 shows a comparison of the thermal conductivity of pure PCM and PCM-graphite matrix for two different PCM. The PCM are a paraffin RT50 and a salt hydrate PCM 72 produced by Merck / Germany. Compared to the pure PCM with a thermal conductivity of 0.2 to 0.5 W/mK, the thermal conductivity of the PCM-graphite matrix is higher by a factor of 50 to 100. This is almost independent of the phase of the PCM, solid or liquid. A comparison where the PCM was solid after production and after cycling shows no reduction of the thermal conductivity. The graphite densities have been 200 g/liter and 175 g/liter.

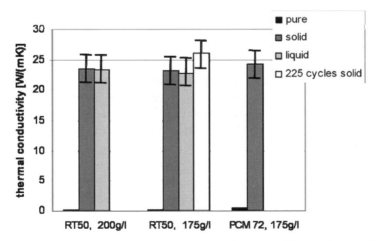

Fig. 2.26. Thermal conductivity of pure PCM, and of PCM-graphite matrix. For the PCM-graphite matrix, values are shown for the PCM being solid or liquid, for different graphite densities, and after 225 cycles (picture: ZAE Bayern).

It is possible to produce the PCM-graphite matrix with many organic and inorganic PCM, however some PCM are difficult to infiltrate into the graphite matrix. Besides this restriction, there is a second disadvantage of the approach with the graphite matrix: there are limitations in producing the graphite matrix with respect to shape and size. Thicknesses larger than 10 mm are difficult to achieve due to the current production process. Therefore, the company SGL TECHNOLOGIES GmbH has developed a second PCM-graphite composite with a different production process.

PCM-graphite compound

In this process, the PCM is mixed with expanded graphite in a compounding process, instead of forming a pre-pressed graphite matrix that is then infiltrated with PCM. The result of the production process is a compound in granular form (f). The advantages of the compound are that it can be produced with any PCM, and that it can be brought into any shape e.g. by injection molding. To do this, the granules are heated above the melting temperature of the PCM; the compound then behaves like a viscous liquid that can be filled into containments of even complex shapes. The compound has a similar volumetric composition as the PCM-graphite matrix, which is about 10 vol.% graphite, 80 vol.% PCM and 10 vol.% air. The thermal conductivity however is lower; about 4 to 5 W/m·K in all directions. The reason for the lower thermal conductivity is that due to the loose contact between the graphite particles in the compound, the graphite particles do not form a connected network as when the graphite is pre-pressed in the case of the graphite matrix. Nevertheless, compared to the pure PCM, the thermal conductivity is still higher by a factor of 5 to 20.

It is often assumed that the thermal conductivity can be calculated similar to the density or the enthalpy of a composite from the volume or mass fractions. However, this is wrong. The fact that PCM-graphite compound and PCM-graphite matrix have very different thermal conductivity, despite having similar graphite content, indicates that the thermal conductivity is not only a function of the volume or mass fractions of the components. To calculate the thermal conductivity the shape, distribution, and orientation of the different components in a composite material must be taken into account.

2.4.3 Encapsulated PCM

Encapsulations are classified according to their size as macro- and microencapsulation, as explained in section 2.3.3. Macroencapsulation is by far the most widely used type of encapsulation, however also microencapsulated PCM are produced on an industrial scale nowadays. When encapsulating PCM, it is necessary to take into account several aspects. First, the material of the container wall must be compatible with the PCM. Then, taking into account the selected wall material, the container wall has to be sufficiently thick to assure the necessary diffusion tightness. Finally, the encapsulation must be designed in a way that it is able to cope with the mechanical stress on the container walls caused by the volume change of the PCM.

2.4.3.1 Examples of macroencapsulation

To encapsulate salt hydrates, usually plastic containers are selected because of material compatibility. Plastics are not corroded by salt hydrates; however, attention has to be paid to the water tightness of the material of the capsule wall. This is to make sure that the water content in the capsule and thus the composition of the salt hydrate does not change with time. Plastic encapsulations can also be used for organic PCM, but the combination of PCM and encapsulation material has to be chosen very carefully because organic materials may soften plastics. Fig.2.27 shows several examples of macro encapsulation in plastic containers from different companies. Many companies produce a selection of different encapsulations. Because plastic containers can nowadays be produced easily in a high variety of shapes, there are few restrictions on the geometry of the encapsulation.

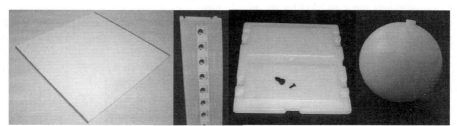

Fig. 2.27. Macroencapsulation in plastic containers. From left to right: bar double panels from Dörken (picture Dörken), panel from PCP (picture: PCP), flat container from Kissmann, and balls from Cristopia, also called nodules.

Another form of macroencapsulation shows fig.2.28. Here, plastic sheets form small containments for the PCM that are sealed with a plastic foil. Such encapsulations called capsule stripes or dimple sheets are useful to cover large surfaces and can be manufactured on a fully automated production line.

Fig. 2.28. Macroencapsulation in capsule stripes as produced by PCP and Dörken for inorganic PCM (picture: ZAE Bayern).

It is of course also possible to use only foils as a wall material, the resulting product is PCM encapsulated in bags. Fig.2.29 shows two examples. To ensure tightness of the bag regarding water, usually plastic foils combined with a metallic layer are used.

Fig. 2.29. Macroencapsulation in bags; left, produced by Climator (picture: Climator), and right, produced by Dörken (picture: Dörken).

If good heat transfer is important, the low thermal conductivity of container walls made of plastic can be a problem. An option is to chose containers with metal walls. Metal walls also have the advantage of higher mechanical stability if a sufficient wall thickness is chosen. It is however necessary to select a suitable metal which is not corroded by the PCM. This selection should also take into account that depending on the metal different options and restrictions for shaping, welding, etc exist. Fig.2.30 shows two examples of metal containers used to encapsulate PCM.

 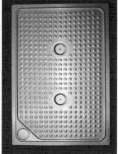

Fig. 2.30. Macroencapsulation in metal containers: left, aluminum profiles with fins for improved heat transfer from Climator (picture: Climator), and right, coated aluminum plate from Rubitherm Technologies GmbH (picture: Rubitherm Technologies GmbH).

On the left is a ClimSel™ Thermal Battery: a hollow aluminum profile with cooling fins that is filled with PCM. The batteries can be connected to each other, as shown in the background. On the right is a coated aluminum plate filled with PCM, which is manufactured by Rubitherm Technologies GmbH.

2.4.3.2 Examples of microencapsulation

Because of the reasons discussed in section 2.3.3, microencapsulation of PCM is technically feasible today only for organic materials. Commercial products seem to use exclusively paraffins. Fig.2.31 shows commercial microencapsulated paraffin, with a typical capsule diameter in the 2-20 μm range, produced by the company BASF.

Fig. 2.31. Electron microscope image of many capsules (picture: FhG-ISE) and an opened microcapsule (picture: BASF).

The microencapsulated PCM is available as fluid dispersion or as dried powder (fig.2.32).

Fig. 2.32. Microencapsulated paraffin produced by BASF, on the left as fluid dispersion, and on the right as dry powder (pictures: BASF).

The microcapsules are sold under the brand name Micronal® as Micronal® DS 5000 (fluid dispersion) and 5001X (dried powder) with a melting temperature of

26 °C, and as Micronal® DS 5007 (fluid dispersion) 5008X (dried powder) with a melting temperature of 23 °C (http://www.micronal.de).

Another source for a commercial product is the company Microtek in the USA, which sells microencapsulated PCM with a wide range of melting temperatures.

2.5 References

[Abhat 1983] Abhat A.: Low temperature latent heat thermal energy storage: heat storage materials. Solar energy, vol. 30, no. 4, 313-332 (1983)

[Atkins et al. 2002] Atkins P.W., de Paula J.: Physical chemistry. Oxford University Press, Oxford (2002) 7th edition, ISBN 0-19-879285-9

[BASF AG] BASF AG. Ludwigshafen, Germany. www.micronal.de

[Bauer et al. 2006] Bauer T., Tamme R., Christ M., Öttinger O.: PCM-graphite composites for high temperature thermal energy storage. Proc. of ECOSTOCK, 10th International Conference on Thermal Energy Storage, Stockton, USA, 2006

[Boese et al. 1999] Boese R., Weiß H.-C., Bläser D.: The Melting Point Alternation in the Short-Chain n-Alkanes: Single-Crystal X-Ray Analyses of Propane at 30 K and of n-Butane to n-Nonane at 90 K. Angew. Chem. Int. Ed. **38**, No. 7, 988-992 (1999)

[Brown et al. 2003] Brown E.N., Kessler M.R., Sottos N.R., White S.R.: In situ poly(urea-formaldehyde) microencapsulation of dicyclopentadiene. J. Microencapsulation **20**, 719-730 (2003)

[Cabeza et al. 2002] Cabeza L., Mehling H., Hiebler S., Ziegler F.: Heat transfer enhancement in water when used as PCM in thermal energy storage. Applied Thermal Engineering **22**, 1141 – 1151 (2002)

[Climator AB] Climator AB. Skovde, Sweden. http://www.climator.com

[CRISTOPIA Energy Systems] CRISTOPIA Energy Systems. Vence, France. http://www.cristopia.com

[Dörken GmbH & Co. KG] Dörken GmbH & Co. KG. Herdecke, Germany. www.doerken.de

[do Couto Aktay et al. 2005] do Couto Aktay K. S., Tamme R., Müller-Steinhagen H.: PCM-Graphite Storage Materials for the Temperature Range 100-300 °C. Proc. of Second Conference on Phase Change Material & Slurry: Scientific Conference & Business Forum, , Yverdon-les-Bains, Switzerland, 15 – 17 June 2005

[Farid et al. 2004] Farid M.M., Khudhair A.M., Razack S.A.K., Al-Hallaj S.: A review on phase change energy storage: materials and applications. Energy Conversion and Management **45**, 1597–1615 (2004)

[Günther et al. 2007] Günther E., Mehling H., Werner M.: Melting and nucleation temperatures of three salt hydrate phase change materials under static pressures up to 800 MPa. J. Phys. D: Appl. Phys. **40**, 4636–4641 (2007)

[Hackeschmidt et al. 2007] Hackeschmidt K, Khelifa N, Girlich D.: Verbesserung der Nutzbaren Wärmeleitung in Latentspeichern durch offenporige Metallschäume, KI Kälte – Luft – Klimatechnik, 33-36 (2007)

[Hafner and Schwarzer 1999] Hafner B., Schwarzer K.: Improvement of the Heat Transfer in a Phase-Change-Material Storage. Presented at 4th Workshop IEA ECES Annex 10 "Phase change materials and chemical reactions for thermal energy storage", Benediktbeuern, Germany, 28.-29. Oktober 1999. www.fskab.com/annex10

[He et al. 1998] He B., Gustafsson M., Setterwall F.: Paraffin Waxes and Their Binary Mixture as Phase Change Materials (PCMs) for Cool Storage in District Cooling System. Presented at 1st Workshop IEA ECES Annex 10 "Phase change materials and chemical reactions for thermal energy storage", Adana, Turkey, 16-17 April 1998. www.fskab.com/annex10

[He et al. 1999] He B., Gustafsson E.M., Setterwall F.: Tetradecane and hexadecane binary mixtures as phase change materials (PCMs) for cool storage in district cooling system. Energy **24**, 1015-1028 (1999)

[Herlach 2004] Herlach D.M.: Solidification and Crystallization. Wiley-VCH Verlag GmbH & Co. KGaA (2004), isbn 3-527-31011-8

[Hiebler and Mehling 2001] Hiebler S., Mehling H.: Latent-Kältespeicherung ohne Eis: Überblick über Materialien und Anwendungen. Proc. of Deutsche Kälte-Klima-Tagung, Ulm, 2001

[Hong and Herling 2006] Hong S.-T., Herling D.R.: Open-cell aluminum foams filled with phase change materials as compact heat sinks. Scripta Materialia, Vol. 55, Issue 10, 887-890 (2006)

[Inaba and Tu 1997] Inaba H., Tu P.: Evaluation of thermophysical characteristics on shape-stabilized paraffin as a solid-liquid phase change material. Heat and Mass Transfer **32**, 307-312 (1997)

[Jahns 1999] Jahns E.: Microencapsulated Phase Change Material. Presented at 4th Workshop IEA ECES Annex 10 "Phase change materials and chemical reactions for thermal energy storage", Benediktbeuern, Germany, 28.-29. Oktober 1999. www.fskab.com/annex10

[Kakiuchi et al. 1998] Kakiuchi H., Yamazaki M., Yabe M., Chihara S., Terunuma Y., Sakata Y.: A study of erythritol as phase change material. Presented at 2nd Workshop IEA ECES Annex 10 "Phase change materials and chemical reactions for thermal energy storage", Sofia, Bulgaria, 11–13 April 1998. www.fskab.com/annex10

[Kashchiev 2000] Kashchiev D.: Nucleation - Basic Theory with Applications. Butterworth Heinemann (2000), isbn 0-7506-4682-9

[Kenisarin and Mahkamov 2007] Kenisarin M., Mahkamov K.: Solar energy storage using phase change materials. Renewable and Sustainable Energy Reviews, Vol.11 (9), 1913-1965 (2007)

[Kurz and Fisher 1992] Kurz W., Fisher D.J.: Fundamentals of Solidification. Trans Tech Publications (1992),3rd edition, isbn 0-87849-522-3

[Lane 1983] Lane G.A.: Solar Heat Storage: Latent Heat Material - Volume I: Background and Scientific Principles. CRC Press, Florida (1983)

[Lane 1986] Lane G.A.: Solar Heat Storage: Latent Heat Material - Volume II: Technology. CRC Press, Florida (1986)

[Lindner 1984] Lindner F.: Latentwärmespeicher - Teil 1: Physikalisch-technische Grundlagen. Brennst.-Wärme-Kraft **36**, Nr. 7-8 (1984)

[Mehling et al. 1999] Mehling H., Hiebler S., Ziegler F.: Latent heat storage using a PCM-graphite composite material: advantages and potential applications. Presented at 4th Workshop IEA ECES Annex 10 "Phase change materials and chemical reactions for thermal energy storage", Benediktbeuern, Germany, 28.-29. Oktober 1999. www.fskab.com/annex10

[Mehling et al. 2000] Mehling H., Hiebler S., Ziegler F.: Latent heat storage using a PCM-graphite composite material. Proc. of TERRASTOCK 2000, Stuttgart, 28.8-1.9.2000

[Mehling 2001] Mehling H.: Latentwärmespeicher – Neue Materialien und Materialkonzepte. Proc. of FVS Workshop 'Wärmespeicherung', DLR Köln, 28th - 29th May 2001. http://www.fv-sonnenenergie.de/Publikationen/index.php?id=5&list=23

[Microtek] Microtek Laboratories. Dayton, USA. http://www.microteklabs.com

[Mills et al. 2006] Mills A., Farid M., Selman J.R., Al-Hallaj S.: Thermal conductivity enhancement of phase change materials using a graphite matrix. Applied Thermal Engineering **26**, 1652–1661 (2006)

[Mutaftschiev 1993] Mutaftschiev B.: Nucleation Theory. In: Hurle D.T.J. (ed.): Handbook of Crystal Growth. Elsevier Science Publishers B.V., 187-247 (1993), ISBN 0-444-89933-2

[Nagano et al. 2000] Nagano K., Mochida T., Iwata K., Hiroyoshi H., Domanski R.: Thermal Performance of Mn(NO$_3$)$_2$·6H$_2$O as a New PCM for Cooling System. Presented at 5th Workshop IEA ECES Annex 10 "Phase change materials and chemical reactions for thermal energy storage", Tsu, Japan, 12-14 April 2000. www.fskab.com/annex10

[Neuschütz 1999] Neuschütz M.: High performance latent heat battery for cars. Presented at 3th Workshop IEA ECES Annex 10 "Phase change materials and chemical reactions for thermal energy storage", Helsinki, Finland, May 1999. www.fskab.com/annex10
[Nikolić et al. 2003] Nikolić R., Marinović-Cincović M., Gadžurić S., Zsigraib I.J.: New materials for solar thermal storage - solid/liquid transitions in fatty acid esters. Solar Energy Materials & Solar Cells **79**, 285–292 (2003)
[Özonur et al. 2006] Özonur Y., Mazman M., Paksoy H. Ö., Evliya H.: Microencapsulation of coco fatty acid mixture for thermal energy storage with phase change material. International Journal of Energy Research **30**, 741-749 (2006)
[Oyama et al. 2005] Oyama H., Shimada W., Ebinuma T., Kamata Y., Takeya S., Uchida T., Nagao J., Narita H.: Phase diagram, latent heat, and specific heat of TBAB semiclathrate hydrate crystals. Fluid Phase Equilibria **234**, 131–135 (2005)
[PCP] Phase Change Products Pty Ltd, short PCP. Perth, Australia. http://www.pcpaustralia.com.au/index.html
[Pimpinelli and Villain 1998] Pimpinelli A., Villain J.: Physics of Crystal Growth. Cambridge University Press (1998), isbn 0-521-55198-6
[Py et al. 2001] Py X., Olives R., Mauran S.: Paraffin / porous-graphite-matrix composite as a high and constant power thermal storage material. International Journal of heat and mass transfer **44**, 2727-2737 (2001)
[Rogerson and Cardoso 2003a] Rogerson M., Cardoso S.: Solidification in Heat Packs: I. Nucleation Rate. AIChE Journal, Vol. 49, No. 2, 505-515 (2003)
[Rogerson and Cardoso 2003b] Rogerson M., Cardoso S.: Solidification in Heat Packs: II. Role of Cavitation. AIChE Journal, Vol. 49, No. 2, 515-521 (2003)
[Rogerson and Cardoso 2003c] Rogerson M., Cardoso S.: Solidification in Heat Packs: III. Metallic Trigger; AIChE Journal, Vol. 49, No. 2, 522-529 (2003)
[Royon et al. 1997] Royon L., Guiffant G., Flaud P.: Investigation of heat transfer in a polymeric phase change material for low level heat. Energy Convers **38**, 517–24 (1997)
[Rubitherm Technologies GmbH] Rubitherm Technologies GmbH. Berlin, Germany. http://www.rubitherm.com
[Sari and Kaygusuz 2003] Sari A., Kaygusuz K.: Some fatty acids used for latent heat storage: thermal stability and corrosion of metals with respect to thermal cycling. Renewable Energy, Volume 28, Issue 6, 939-948 (2003)
[Satzger et al. 1997] Satzger P., Eska B, Ziegler F.: Matrix-heat-exchanger for latent-heat coldstorage; Proc. of 7th International conference on thermal energy storage (1997)
[Schröder 1985] Schröder J.: Some Materials and Measures to Store Latent Heat. Proc. of IEA Workshop on "Latent heat stores – Technology and applications", Stuttgart, 1985
[SGL Technologies GmbH] SGL TECHNOLOGIES GmbH. Meitingen, Germany. www.sglcarbon.com/eg
[Sharma et al. 2004] Sharma S.D., Kitano H., Sagara K.: Phase change materials for low temperature solar thermal applications. Res. Rep. Fac. Eng. Mie Univ., Vol. 29, 31-64 (2004)
[Steiner et al. 1980] Steiner D., Heine D., Heess F.: Untersuchung von Mittel- und Hochtemperatur Latentwärmespeicher Materialien. Schlussbericht BMFT ET 4335 (1980)
[Tunçbilek et al. 2005] Tunçbilek K., Sari A., Tarhan S., Ergüneş G., Kaygusuz K.: Lauric and palmitic acids eutectic mixture as latent heat storage material for low temperature heating applications. Energy **30**, 677–692 (2005)
[Velraj et al. 1999] Velraj R., Seeniraj R.V., Hafner B., Faber C., Schwarzer K.: Heat transfer enhancement in a latent heat storage system. Solar Energy, vol. 65, No. 3, 171–180 (1999)
[Voigt 1993] Voigt W.: Calculation of salt activities in molten salt hydrates applying the modified BET equation, I: Binary systems. Monatshefte für Chemie **124**, 839-848 (1993)
[Xiao et al. 2006] Xiao R., Wu S., Tang L., Huang C., Feng Z.: Experimental investgaton of the pressure drop of clathrate hydrate slurry (CHS) flow of Tetra Butyl Ammonium Bromide (TBAB) in straight pipe. Proc. of ECOSTOCK, 10th International Conference on Thermal Energy Storage, Stockton, USA, 2006

[Yinping et al. 2006] Zhang Yinping, Zhou Guobing, Yang Ruib Lin Kunpinga: Our Research on Shape-stabilized PCM in Energy-efficient Buildings; Proc. of ECOSTOCK, 10th International Conference on Thermal Energy Storage, Stockton, USA, 2006

[Zalba et al. 2003] Zalba B., Marin J.M., Cabeza L.F., Mehling H.: Review on thermal energy storage with phase change materials, heat transfer analysis and applications. Applied thermal Engineering **23**, 251 – 283 (2003)

3 Determination of physical and technical properties

The most important physical and technical properties of phase change materials have been discussed in section 2.1. Several of these properties also apply to whole objects like macroencapsulated PCM; the most important one is the capacity to store heat in a small temperature range. The physical and technical properties are the basis for the development and design of any product. Therefore, their correct determination and the knowledge of their accuracy are essential. This is the topic of this chapter.

3.1 Definition of material and object properties

Before discussing physical and technical properties, it is necessary to define what a material and an object is, and what their respective properties are. This is not always clear to non-scientists and can lead to misunderstandings. In this chapter, a clear definition is crucial.

The word material can have many meanings. Here the definition of a *material* is: the substance out of which a thing is made. A material is something that has therefore homogeneous properties throughout, as fig.3.1 shows. That means if a sample is taken out of a large piece of the material and a *material property* is measured, the value will be the same independent of where that sample has been taken and also independent of the size and geometry of the sample. An example for a material property is the thermal conductivity. From its definition, it does not depend on the size or shape of the sample. Copper is a material and therefore the thermal conductivity of copper does not depend on the size of the sample or its geometry. If copper powder is mixed into a plastic material and the mixture is homogeneous, the result is a *composite material*.

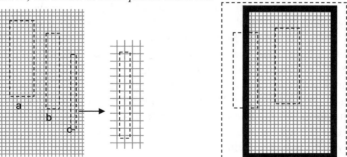

Fig. 3.1. Materials are homogeneous at some scale (left) and thus show the same property called material property. Objects (right) are not homogeneous.

Materials are homogeneous with reference to a property at some scale, and thus show the same property called material property. If the size of the sample is decreased to the point where the sample becomes non-homogeneous, the measured value is not a material property any more but only reflects the property of that specific sample. This means that the relation between the "structure of the material" and the "size of the sample" determines if a material property is measured or not. The left side of fig.3.1 shows a thermally conducting grid. Parts a) and b) are significantly larger than the structure of the grid and the thermal conductivity is a material property. Part c) is too thin; it is not homogeneous any more as the structure of the grid is similar to the thickness. If a "thermal conductivity" of such a sample is determined using a standard method, the experimental result is not the same as for part a) or b). In that case, the measured value is often called an "apparent" thermal conductivity. The same holds for composite materials: a material property can only be measured if the sample is large enough to be homogeneous.

On a large scale, non-homogeneous properties arise due to a coating, encapsulation, or if the measured property also depends on the geometry. *Objects*, as shown on the right side of fig.3.1, are not homogeneous. Therefore, the *object property* is a property of the whole object, and consequently has to be measured on the whole object. Optionally the object property can also be calculated from the material properties of the components and the object structure and geometry. For example, encapsulated PCM like the PCM in the metal box with fins in fig.2.30 have the object properties mass and weight, but not the material property density. The density is different if a sample from the centre is taken, where there is only PCM, or from the surface where there is also material of the encapsulation. There is also an influence from ambient conditions on some properties. If the box with fins releases heat to the environment, the heat released and the heat transfer coefficient depend on the ambient conditions and have to be determined under defined ambient conditions. They are therefore object properties under defined ambient conditions, similar as C_P and C_V.

The specific sample investigated, and the property that is determined, have a strong influence on the selection of the measurement system and how a measurement has to be performed. Until recently, no standard existed for the determination of the physical and technical properties of PCM and the presentation of data in graphs or tables. Because PCM have an extraordinary high heat storage density in a small temperature range, they pose special problems for example in calorimetry. Consequently, common standards in calorimetry for materials with no or little phase change enthalpy cannot be applied to PCM. To solve this problem several companies contracted the ZAE Bayern and FhG-ISE. The new standard in the currently valid form is available at http://www.pcm-ral.de. Here, some of the basic problems and the approach to solve them in the new standard are explained. For the density and viscosity, useful standard methods exist and these topics are therefore not discussed here.

3.2 Stored heat of materials

The *stored heat* between two temperatures is the most important characteristic of PCM materials and objects, as the final goal in an application is to store heat. Before starting with the detailed discussion, it is necessary to go through some definitions.

3.2.1 Basics of calorimetry

The stored heat can be given with respect to the volume or mass in case of a material, or with respect to a given object. The *calorimetric formula*

$$dQ = \frac{dQ}{dT} \cdot dT = C \cdot dT \tag{3.1}$$

defines the relation between the stored heat and the heat capacity C (e.g. J/K). The heat capacity of a material can be given with respect to mass (e.g. J/kg·K), volume (e.g. J/m³·K) or amount of the material (e.g. J/mol·K). If the heat capacity is with respect to mass, volume, or amount, it is called specific heat capacity and c is used instead of C. The heat capacity depends also on the boundary conditions; for constant pressure, it is denoted as C_p, and for constant volume as C_v.

Because of the relation between the enthalpy H, internal energy U, pressure p, and volume V

$$H = U + pV \tag{3.2}$$

and the *first law of thermodynamics*

$$dU = dQ - pdV \tag{3.3}$$

follows

$$dH = dU + d(p \cdot V) = dQ - pdV + pdV + Vdp = dQ + Vdp. \tag{3.4}$$

For constant pressure follows then

$$dH = dQ \tag{3.5}$$

and consequently also

$$\Delta H = \Delta Q. \tag{3.6}$$

Further on

$$\left.\frac{dH}{dT}\right|_p = \left.\frac{dQ}{dT}\right|_p = C_p \tag{3.7}$$

For solids and liquids, which under usual conditions are incompressible, the term $V \cdot dp$ can be neglected in eq.3.4. Then, independent of the pressure conditions, eq.3.5, eq.3.6, and eq.3.7 are valid. Changes of enthalpy Δ H and of heat Δ Q are therefore often used synonymously when dealing with solids and liquids. From a physics point of view, there is however an important difference between H and Q: heat is something that only shows up in the transfer of thermal energy and the amount transferred depends on the path between the initial and the final state of the system. Enthalpy H is a thermodynamic potential and therefore characteristic of a state. The correct expression is therefore that the heat stored or lost is equal to the enthalpy increase or reduction of the storage material.

Methods to determine the change of heat in any kind of process are called *calorimetric methods* (from Latin calor = heat). A *calorimeter* usually works in one of the following ways shown in fig.3.2, based on the calorimetric formula (eq.3.1):

- The calorimeter actively supplies heat to the sample via an electrical resistance heater and measures the temperature change of the sample.

  The electrical energy supplied is calculated from the voltage and current of the heater.

- The calorimeter actively supplies heat to the sample via a thermal resistance and a temperature difference between the sample and its surrounding. Further on, it measures the temperature change of the sample.

  The heat supplied to the sample across the thermal resistance is calculated from the temperature gradient across the thermal resistance. The value of the resistance can be calculated. However, usually it is determined via a calibration using standard materials with a known heat capacity or melting enthalpy.

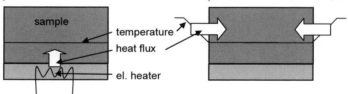

Fig. 3.2. Supply of heat electrically (left) or by a heat flux from the outside (right).

An advantage of the electrical heating is that it needs no calibration. On the other hand, it can only be used for heating, while with a thermal resistance there is also the possibility of measuring upon cooling when the temperature of the surrounding is below the sample temperature.

There are several *modes to operate calorimeters* to determine the heat storage capacity within a temperature interval of interest. They differ in the resolution of the data on the heat supplied to the sample and the corresponding temperature change:

1. *Interval mode*: heating or cooling is between the start and the end of the temperature interval of interest, indicated on the left in fig.3.3 by "|". Then

$$\Delta Q = \int_{T_{start}}^{T_{end}} C_p dT = C_p \cdot \Delta T = C_p \cdot (T_{end} - T_{start}). \tag{3.8}$$

It is important that the sample is isothermal at the start and end temperatures, otherwise these states are not well defined and the heat supplied is not determined correctly. The heat supplied or extracted for the temperature change can be measured with high accuracy; the temperature resolution of the stored heat however is very poor. The heat capacity value determined is an average value for the whole interval. Fig.3.3 shows this for the case of a material with a melting temperature, and not a melting range.

2. *Step mode*: heating or cooling is as with the interval mode, but the whole interval of interest is now divided into small intervals, called steps. For these steps, the same treatment as described for the interval mode before is applied. This increases the temperature resolution of the stored heat. The resolution is equal to the height of the temperature steps, however the start and end of each step must again be an isothermal state, otherwise the change in heat in each step is not determined correctly.
3. *Dynamic mode*: the interval is scanned by continuously heating or cooling according

$$dQ = \frac{dQ}{dT} \cdot dT = C_p \cdot dT. \tag{3.9}$$

There are no steps to wait for thermodynamic equilibrium any more. The heating or cooling rate has to be slow enough to assure thermodynamic equilibrium within the sample however. If not, the heat supplied at each data recording (equal to the enthalpy change dH, eq.3.5) cannot be assigned to the measured temperature. Too high heating or cooling rates cause significant errors in the data of the heat stored as a function of temperature. The resolution of the data itself is only restricted by the data recording system.

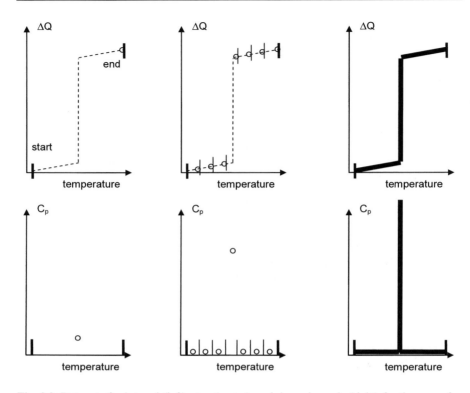

Fig. 3.3. Data sets for interval (left), step (center), and dynamic mode (right) for the case of a material with a melting temperature. The total heat ΔQ supplied to the sample ($= \Delta H$) over the whole interval is the same, but the resolution of the data ○ increases from left to right.

From the data of the heat supplied to the sample and the corresponding temperature change the heat capacity can be calculated. Actually, in the case of a melting temperature the heat capacity, which is the 1st derivative of the stored heat with respect to temperature, would be infinite. Therefore, many people do not like to call the values a heat capacity, even if there is a melting range instead of a melting temperature. However, for a melting range it is not possible to distinguish the temperature ranges of melting and of only sensible heat storage. Any attempt to do it anyway leads to confusion and uncertainties as the past has shown. Further on, even for a melting temperature the measurement will never result in an infinite value. Years of experience have shown that the best way is to strictly follow the definition and use the heat capacity throughout, even in the presence of a melting temperature.

Fig.3.3 shows data sets for interval (left), step (centre) and dynamic mode (right) for the case of a melting temperature. The total heat supplied to the sample over the whole interval results from adding up the heat supplied in each interval. As the top row of diagrams shows, it is the same for all measurement modes, but the resolution of the data o increases from left to right. The influence of the resolution on the heat capacity data shown in the lower row of diagrams is dramatic. It is obvious that the interval mode gives the worst resolution because it averages over the whole measurement interval where, due to the phase change, dramatic differences in the heat capacity occur. Between step and dynamic mode, there is no significant difference in the range of sensible heat storage. In the range of phase change, the width of the measurement interval has a strong influence on the peak value of the heat capacity data. This effect can be dramatic when pure substances with a melting temperature are analyzed, but it is usually less pronounced when measuring PCM because most PCM have a melting range.

General aspects in doing a measurement

From what was discussed up to this point in chapter 3, it is possible to summarize that aspects related to doing a measurement include general aspects of calorimetry and the materials investigated. Three main aspects have to be considered:

1. The sample has to be representative for the investigated material

   This originates from the definition of a material property (section 3.1).
   To assure that the sample is representative, the sample must be larger than typical inhomogeneities of the material.

2. Correct determination of the exchanged heat and temperature of the sample

   This originates from the use of the calorimetric formula.
   For the temperature, the correct determination is usually assured by a calibration procedure, where standard materials with known melting temperatures are used. This procedure has to be repeated from time to time. Usually the temperature is not measured within, but only close to the sample. This can lead to an offset between the measured and the real sample temperature.
   For the heat flow, the correct determination is usually also assured by a calibration. If a measurement has to be very accurate, the calibration should be done just before or after a sample measurement.

3. Thermodynamic equilibrium in the sample

   Thermodynamic equilibrium in the sample requires first that *the sample is isothermal*, otherwise the heat flux cannot be attributed to a single temperature indicated at the sensor. Therefore the measurement should be done slow enough, or the sample should not be too large. This can be in conflict with the necessity to have a sufficiently large sample to be representative for the investigated material.
   Further on, the sample should be in *reaction equilibrium*, otherwise the enthalpy at the measured temperature has no defined value, and consequently enthalpy differences are also not well defined. This refers to two aspects: first,

dynamic processes have to proceed to a stable state, and second, there should only be one stable state at the same temperature. There are several occasions causing problems with regard to reaction equilibrium, for example very slow reactions, metastable states like amorphous instead of crystalline structures, subcooling, and different crystalline structures.

3.2.2 Problems in doing measurements on PCM

Besides the general aspects just summarized that apply to any material, PCM pose additional problems or larger problems due to their high phase change enthalpy.

Regarding point 1, saying that the sample has to be representative for the investigated material, subcooling can be a problem. Subcooling can depend on the sample size and therefore the sample size should be typical for the application if possible.

Regarding point 2, the correct determination of the exchanged heat and temperature of the sample, PCM will give a large signal in the temperature range where the sample melts, and a small signal outside that range. The calibration therefore should be done for large signals.

Regarding point 3, the thermodynamic equilibrium in the sample, the measurement on PCM can cause severe problems. The requirement of reaction equilibrium is violated during subcooling of a sample, or if upon cooling crystals form slowly. This means the heating / cooling rate should be small. To get the sample isothermal can also take a lot of time with a PCM during the phase change, because of the large amounts of heat involved and the usually low thermal conductivity of PCM. If not done properly, there will be significant errors as fig.3.4 shows. When the temperature of the sample is measured at the sample surface, which is usually the case, the heat released or supplied to the sample is then attributed to a wrong temperature, too low when cooling and too high when heating.

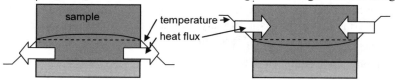

Fig. 3.4. Left: temperature profile in a cooling measurement for a material storing only sensible heat (- - -) and a PCM during phase change (—). Right: the same for a heating measurement.

Outside the melting temperature range, this effect is relatively small; because only sensible heat is stored, the temperature gradients remain small. While most calorimeters are constructed to give accurate results for sensible heats of "ordinary" materials, a significant offset can be observed in the melting range of PCM with their special, high phase change enthalpy. Upon cooling, the situation can even be worse. In most cases, interval and step mode are not critical, as isothermal conditions are attained when waited for long enough. In dynamic mode, where also many data points lie in the melting and solidification temperature range of the PCM, different measurement results are obtained on heating and cooling (fig.3.5, left). This is often called hysteresis; however, most people understand *hysteresis* as a property of the material / sample only. Effects, which are only due to the measurement conditions, are usually not called hysteresis, but *apparent hysteresis*.

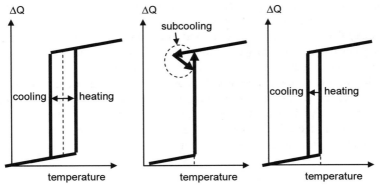

Fig. 3.5. Left: apparent hysteresis caused by non-isothermal conditions. Center: real hysteresis as a material property caused by subcooling. Right: real hysteresis by slow heat release or a real difference between the phase change temperatures.

To find out if an observed hysteresis is real or only apparent, that means if it is a property of the material or only due to the measurement apparatus or procedure, it is necessary to have isothermal conditions in the sample. Therefore, the heating and cooling rate should be small or the sample size should be small. However, small sample size contradicts the requirement of having a representative sample, and small heating and cooling rates will lead to a low signal to noise ratio. Therefore, it often cannot be avoided to have different results from heating and cooling measurements. To get at least an idea what the result would be if temperature equilibrium were attained, it is necessary to measure always upon heating and cooling and with a small heating and cooling rate as possible. The closer the result, the better temperature equilibrium was attained. A comparison of results from measurements with different heating rates indicates also how close temperature equilibrium was attained.

There are several effects due to the material, which cause real hysteresis. The most common one is subcooling. Subcooling will lead to different results for heating and cooling; however, it is not a shift of the results to higher temperatures on heating and lower temperatures on cooling. Subcooling is restricted to the temperature range where solidification is started, as fig.3.5 shows. Another cause of hysteresis is when the latent heat is released too slowly on cooling, e.g. because the crystal lattice forms very slowly or because diffusion processes are necessary to homogenize the sample. The temperature of the sample then drops below the melting temperature. Upon melting, a similar effect is usually not observed (fig.3.5, right) because during melting the kinetic effects proceed much faster. Another possible cause of hysteresis is when upon solidification the sample forms a different solid phase than the one it was at the beginning of melting process. This can be an amorphous phase, or a different crystalline structure.

Up to now, the discussion and the diagrams discussed the case of a melting temperature. Because PCM are usually not pure substances and therefore usually have a melting range, it is not possible to detect from any single measurement what exactly the underlying effects are and to eliminate the deviations in the data caused by not having isothermal conditions, reaction equilibrium, etc. It is therefore strongly recommend to try to measure at least as close as possible to isothermal conditions and to make measurements on heating and additionally on cooling. When discussing calorimeter types and working principles in section 3.2.4, this topic will be discussed again. The problems in doing measurements on PCM, discussed here on a theoretical basis, will then be discussed in connection with the measurement systems and real measurement results.

3.2.3 Problems in presenting data on PCM

In a temperature range where a material does not undergo a phase change it only stores sensible heat and its capacity to store heat is characterized by only one number: the heat capacity. The heat capacity is usually almost constant within a temperature interval in the order of 10 K. When a material undergoes a phase change, the situation is more complex and usually described by several numbers. Regarding PCM, four values are usually tabulated:

- phase change temperature
- phase change enthalpy
- heat capacity in the solid state
- heat capacity in the liquid state

These properties however are well defined only for pure materials like water with a sharp melting temperature. Most PCM however are not pure materials and therefore instead of a melting temperature there is a melting range. An example shows fig.3.12. It is therefore not possible to separate sensible and latent heat in measured data. Until recently, it was common to take the onset temperature or peak temperature of the melting curve in a measurement as the melting temperature, but as the case in fig.3.12 shows, this does not reflect the reality at all. In such cases, this practice leads to misinterpretations. If a PCM has a melting range of several K, but the data tables give a sharp melting temperature, this can lead to significant errors in the design of a system.

The way out of such a problem is often to go back to the original goal: for calorimetry, it is the description of heat storage (Mehling et al. 2006). What is necessary is the stored heat as a function of temperature. The stored heat as a function of temperature only refers to the actual problem of heat storage and heat transfer; it avoids the definition of a single melting temperature, a constant heat capacity that might not exist, as well as the separation between sensible heat and latent heat. Therefore, it is recommended to give the *stored heat as a function of temperature in given temperature intervals*. This is nearly identical with the common definition of the heat capacity as the heat stored per temperature interval of 1 K. The term "stored heat as a function of temperature in given temperature intervals" avoids ambiguities of the definition of the melting range and temperature. By avoiding the term "heat capacity", the conflict in the separation between latent and sensible heat is also resolved. This could also have been achieved using the "enthalpy as a function of temperature". The meaning of this term is clear to scientists, however not to many other people interested in PCM technology. The "stored heat as a function of temperature in given temperature intervals" can be calculated from the stored heat or the enthalpy change, as shown in fig.3.6.

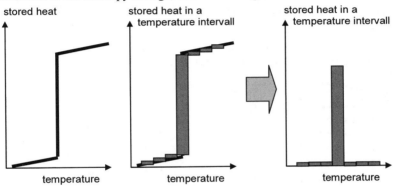

Fig. 3.6. Determination of the "stored heat as a function of temperature in given temperature intervals".

If the stored heat is given with respect to sample mass, volume, or amount, the result is the heat capacity of the material; this includes latent and sensible heat. For the goal to describe the storage of heat, it is finally not important which fraction of the heat is stored as sensible or as latent heat. To calculate the stored heat per volume it is advisable to multiply the stored heat per mass with the minimum density in the temperature range of operation.

The effect of hysteresis is another problem in presenting data (fig.3.5, right). In the case of hysteresis, there are different data from heating and cooling measurements. Hysteresis can be an artifact of the measurement procedure when the heating or cooling rate is too high, but it can also be a real property of the material. Therefore, it is not correct to choose only the data from the heating or from the cooling experiment. To resolve this problem, it is recommended to give the *stored heat as a function of temperature in given temperature intervals for the heating as well as for the cooling experiment.* Most of the data available in current literature are not in this form. The reader therefore has to be aware of the potential problems described above, when using literature data.

Finally, there is one problem left in presenting data. If in the cooling experiment the sample subcools (fig.3.5, center), how is the "stored heat as a function of temperature in given temperature intervals" calculated? Subcooling is quite common and means that there are different states at the same temperature, at least in some temperature range. Therefore, it is impossible to give the stored heat as one value in a temperature interval, even when heating and cooling are treated separately. In addition, subcooling depends on many things, including the sample size, and is therefore not predictable. Usually there is a temperature range where crystallization starts. The only way out of this dilemma is to eliminate the effect of subcooling from the data of the "stored heat as a function of temperature in given temperature intervals" and to give the degree of subcooling as separate information. The stored heat, which is lost upon subcooling, is attributed to the solidification temperature, because for very large sample and / or no subcooling, this is what it would be. Fig.3.7 describes the procedure.

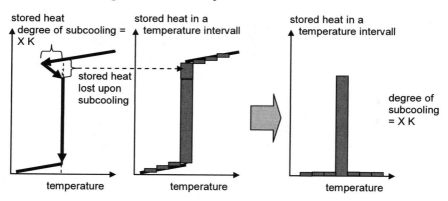

Fig. 3.7. Determination of the "stored heat as a function of temperature in given temperature intervals" plus "degree of subcooling".

Therefore, the effect of subcooling should be eliminated from H-T-graphs (or stored heat as a function of temperature) and be given separately with reference to the sample mass and other conditions that have an influence. If the phase change temperature is not clearly defined as a reference temperature to calculate the degree of subcooling, the temperature where crystallization starts should be given.

It is also possible to tabulate the stored heat as a function of temperature. This considerably simplifies reading the data for people without a scientific background. From the tables it is easy to calculate the heat stored in any arbitrary temperature interval.

Because this standard is rather new, very few data are currently available following it and consequently most data in this book are still given as melting temperatures and melting enthalpies.

3.2.4 Calorimeter types and working principles

After discussing general aspects of calorimetry, and problems in doing measurements on PCM and presenting data, it is now time to turn to the description of calorimetric methods, measurement and evaluation procedures. There is a broad range of different methods with distinct advantages and disadvantages. The most common ones used to make calorimetric measurements on PCM will be discussed in detail: differential scanning calorimetry (DSC) in dynamic mode, in step mode, and T-history method. For more information on calorimetry the excellent books by Speyer 1994 and Hemminger and Cammenga 1989 are recommended. Höhne 1990, Cammenga et al. 1993, Schubnell 2000, Gmelin and Sarge 1995, and Gmelin and Sarge 2000 describe the topic of calibration with respect to temperature and heat flow in detail; a topic which is only briefly mentioned here. A study on the accuracy of the different methods by an intercomparison test on a paraffin and a salt hydrate is described in Mehling et al. 2006.

3.2.4.1 Differential scanning calorimetry in dynamic mode

Differential scanning calorimetry (DSC) is a form of calorimetry where a temperature range is scanned. In addition to the sample, also a reference is subject to the same temperature program. The reference is used to determine the heat stored in the sample by the difference in the signal of the sample compared to the reference. The advantage of this is a higher precision of the determination of the heat flow into the sample.

DSCs can be constructed based on the two working principles shown in fig.3.2, namely as power compensating calorimeters and as heat exchanging calorimeters. *Power compensating* calorimeters are not very common and therefore not described here in detail. They use electrical heaters to compensate the temperature difference between sample and reference when a temperature range is scanned. *Heat exchanging calorimeters*, which are also called *heat flux DSC (hf-DSC)*, are very common. They determine the difference in the heat exchanged between ambient and sample and between ambient and reference, via a thermal resistance R_{th}. The thermal resistance in a hf-DSC can be constructed as disk-type, where the heat flux is realized and measured through a disc shaped support, or as cylinder-type, where this is done at the surface of the cylindrical sample and reference. The cylinder-type hf-DSC just has a different design for the heat-conducting path and for the measurement of temperatures.

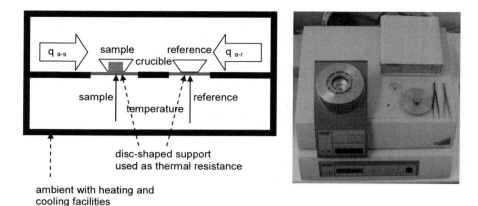

Fig. 3.8. Schematic design of a disc-type hf-DSC and commercial hf-DSC (picture: ZAE Bayern)

Fig.3.8 shows the typical design of a disc-type system with a sample and a reference crucible. To supply or extract heat from the sample, it is necessary to change the temperature of the ambient of the sample with time. In the heating mode, this is done with an electrically heated furnace; in the cooling mode, it is done with a compression cooler or liquid nitrogen. Commonly, the measurements are performed in the dynamic mode, which means the ambient is heated or cooled at a constant rate. While the temperature of the ambient is changed, the heat flux to the sample, as a reaction to the temperature difference between sample and ambient, is measured.

For a quantitative evaluation, a series of measurements is necessary. In these measurements, the reference crucible is typically empty, as shown in fig.3.8. The task of the reference crucible is to eliminate the thermal effect of the sample crucible and heat losses, so that the final signal is mainly due to the sample. In other words, the difference setup increases the accuracy of the measurement of the heat flow to the sample by eliminating the effect of the crucible. The following derivation of the mathematical treatment will explain this. In this derivation, sample side (s) and reference side (r) refer to the respective crucible and its content, unless stated otherwise. Due to the design of the DSC, sample side and reference side have a symmetric heat conducting connection (support) to the ambient such that the heat flux between the sample side and the reference side to the ambient (a) are proportional to the temperature difference

$$\dot{Q}_{a-s} = \frac{1}{R_{th}} \cdot (T_a - T_s) \text{ and } \dot{Q}_{a-r} = \frac{1}{R_{th}} \cdot (T_a - T_r). \tag{3.10}$$

The difference of both heat fluxes is

$$\dot{Q}_{a-s} - \dot{Q}_{a-r} = \frac{1}{R_{th}} \cdot (T_a - T_s) - \frac{1}{R_{th}} \cdot (T_a - T_r) = \frac{1}{R_{th}} \cdot (T_r - T_s). \qquad (3.11)$$

This means the difference in the heat flux is simply proportional to the difference in the temperatures of the sample and the reference side; the ambient temperature cancels out. The sensitivity of a measurement instrument is the ratio of the measured signal, here the temperature difference, to the value to be determined, here the heat flux. Generally holds that a higher sensitivity means a larger signal. According to eq.3.10, this means the *sensitivity* of the DSC in dynamic mode is the ratio of the temperature difference to the heat flow. The sensitivity is therefore equal to the thermal resistance R_{th} of the conducting connection; it is thus a property of the system design and not the measurement parameters. The unit of R_{th}, and therefore of the sensitivity, is K/W; however, often also mK/mW or μV/mW is used when referring to the voltage difference of the thermocouples instead of the temperature.

Assuming that the crucibles and their content are isothermal, it follows that

$$\dot{Q}_{a-s} = \frac{d}{dt} Q_s = \frac{dT_s}{dt} \cdot C_s \text{ and } \dot{Q}_{a-r} = \frac{d}{dt} Q_r = \frac{dT_r}{dt} \cdot C_r. \qquad (3.12)$$

If only sensible heat is stored with little change in the heat capacity, the heating rate will be equal to the heating rate of the ambient ß

$$\beta = \frac{dT_s}{dt} = \frac{dT_r}{dt} = \frac{dT_a}{dt}. \qquad (3.13)$$

To simplify the evaluation the standard evaluation procedure assumes that this is in general the case. This is however wrong during the melting process and can lead to significant errors in the result of c_p and h as a function of temperature (Hiebler 2007).

From eq.3.11, eq.3.12, and eq.3.13 follows that

$$\dot{Q}_{a-s} - \dot{Q}_{a-r} = \frac{1}{R_{th}} \cdot (T_r - T_s) = \beta \cdot (C_s - C_r). \qquad (3.14)$$

Therefore, the measured temperature difference is proportional to the heat capacities of sample and reference

$$\Delta T = T_r - T_s = \beta \cdot R_{th} \cdot (C_s - C_r). \qquad (3.15)$$

This is the basic *equation to evaluate measurements* using DSC in the dynamic mode. The ratio of the measured signal ΔT to the determined property C is the proportionality constant $\beta \cdot R_{th}$, that is the heating rate multiplied with the sensitivity of the calorimeter. Therefore, measurements with small heating or cooling rates will result in low signal to noise ratio for heat capacity data.

To evaluate a measurement, the value of R_{th} needs to be known. As the system does not give an absolute signal, this is done by a calibration. For calibration, materials with a well-defined thermal effect are used; that can be the heat capacity or the enthalpy of a phase change. From the known thermal effect and the corresponding measurement signal, the sensitivity can be calculated using eq.3.15. Depending on the calibration chosen, heat capacity or enthalpy, different measurement procedures can be applied. Actually more often, the desired result determines the measurement procedure, which in turn uses a given calibration.

Standard measurement and evaluation procedure using heat capacity calibration

A measurement of the heat capacity of a sample based on a calibration with a c_p-standard comprises three measurements with the system. The reference crucible is empty in all these measurements. Fig.3.9 shows schematically the temperature-time histories and recorded signals for these three measurements for heating and a sample that melts at a defined melting temperature. Usually a DSC records the thermocouple voltage difference between sample and reference. Therefore, the sensitivity of the system is often given in µV/mW instead of mK/mW.

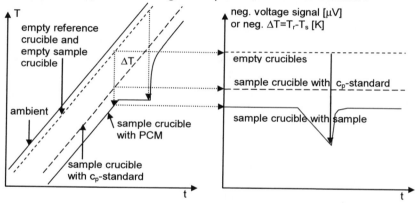

Fig. 3.9. Temperatures and measured signals (temperature differences) in dynamic mode in a heating experiment. The sample has a defined melting temperature.

Fig.3.10 shows the signal from a real measurement on a PCM. Compared to the peak in fig.3.9, the software automatically changes the sign of the signal in a way that the endothermic effect of the sample upon heating leads to an upward peak.

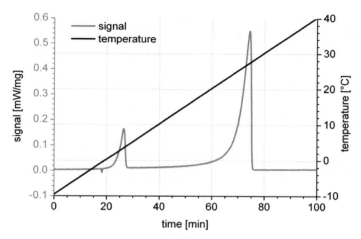

Fig. 3.10. Typical heating ramp and signal during a DSC measurement with the dynamic method (picture: ZAE Bayern)

First, a measurement with empty reference and empty sample crucible is performed (fig.3.9) giving

$$\Delta T_{empty} = \beta \cdot R_{th} \cdot [C_{s,crucible} - C_{r,crucible}]. \tag{3.16}$$

This takes into account the effect that the sample and reference crucibles never match exactly and the system is never fully symmetric. Then a c_p-standard, made for example from sapphire or copper, is put into the sample crucible and the measurement is repeated (fig.3.9)

$$\Delta T_{standard} = \beta \cdot R_{th} \cdot [(C_{s,crucible} + C_{s,standard}) - C_{r,crucible}]. \tag{3.17}$$

This is used to determine the unknown proportionality constant $\beta \cdot R_{th}$ to calibrate the system. The same is then repeated with the sample in the sample crucible (fig.3.9)

$$\Delta T_{sample} = \beta \cdot R_{th} \cdot [(C_{s,crucible} + C_{s,sample}) - C_{r,crucible}]. \tag{3.18}$$

In the evaluation, the effect of the crucibles is eliminated by taking the differences between standard and sample data to the empty data. The calibration is taking into account by taking the sample relative to the standard signal. Taking everything together in one equation gives

$$\frac{\Delta T_{sample} - \Delta T_{empty}}{\Delta T_{standard} - \Delta T_{empty}} = \frac{C_{sample}}{C_{standard}} = \frac{c_{p,sample} \cdot m_{sample}}{c_{p,standard} \cdot m_{standard}}. \tag{3.19}$$

From the known mass of sample and standard, and the specific heat capacity of the standard, the specific heat capacity of the sample can be calculated at any point of time in fig.3.11. Because in a heating measurement a larger heat capacity will result in a lower temperature recording, the downward peak in the recorded temperature results in an upward peak in the heat capacity data.

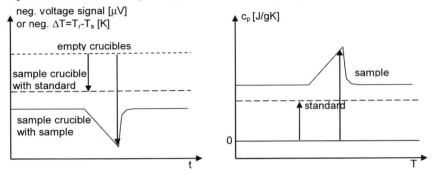

Fig.3.11. Measured signals at different times (left), heat capacity of the standard and calculated heat capacity of the sample (right).

The final step, the conversion between time and temperature in fig.3.11, is done with another calibration. For this, standard materials with different melting temperatures are measured at the same heating rate. Using the known melting temperatures of these standard materials, the temperature offset between the ambient, which is always recorded, and the sample crucible can be calibrated.

At this point, where the discussion of the measurement and evaluation procedure for DSC in dynamic mode and with heat capacity calibration is completed, there are some important points that have to be mentioned. First, in the case of the heat capacity calibration, the sensitivity cancels out in eq.3.19 due to the mathematical evaluation procedure. Consequently, the sensitivity is never calculated directly and never shows up as a value in an evaluation. Second, it is important to note that in fig.3.9 an ideal example is discussed, where the sample has a sharp melting temperature and is isothermal during melting. This is why the sample crucible has a constant temperature for some time when the melting temperature is reached. In fig.3.11 on the right, the evaluation result seems to indicate a melting range. This is because the reference or the ambient temperature is usually used for the temperature axes and not the sample temperature, and it is changing even while the sample temperature stays constant. The mechanism becomes clear when following the transition from the T-t curves on the left of fig.3.9 to the signal-t curves on the right. To avoid this effect it would be necessary to use the sample temperature for the temperature axes. This however would cause even more problems. If the sample temperature is used on a cooling measurement, the sample temperature could rise due to crystallization of a subcooled sample and the resulting curve would change its direction and go from right to left. This is avoided when using the reference temperature for the axes. The fact that even a sharp melting temperature will show up as a melting range in the signal has another important consequence: in a cooling measurement, the peak of the signal will show up at a lower temperature and in heating at a higher one. The real melting temperature is somewhere in between. This means, even if the sample is isothermal during the measurement there will be apparent hysteresis in the measurement result! This is important for the interpretation of measured data. When using metallic standard materials like indium with very high thermal conductivity, which are isothermal during a measurement, this effect is seen in the finite peak width in the measurement result instead of a sharp melting temperature. Third, due to the measurement with constant heating rate, there is also no thermal equilibrium in the sample when measuring a PCM. This can lead to deviations of several K of the indicated heat storage capacity with respect to temperature (Schneker and Stäger 1997). Experimentally, the deviation can be reduced by using smaller heating and cooling rates or smaller sample mass. Fig.3.12 shows results of measurements that were done on the same material but with different sample mass and heating rates.

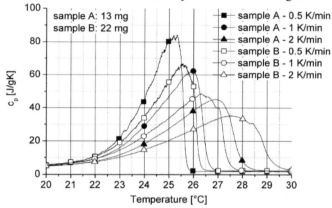

Fig. 3.12. Measurement in dynamic mode on two samples of the same material but with different mass and heating rate. The symbols mark different curves (picture: ZAE Bayern).

The end of the phase change peak in the $c_p(T)$-curve is shifted from 26 °C (0.5 K/min, small mass) to about 30 °C (2 K/min, large mass). A reduction of the deviation is only possible at the expense of a weaker signal/noise ratio, which is clearly visible in fig.3.12.

After this excursion, the next possibility of measurement and evaluation procedures for hf-DSC is now discussed.

Standard measurement and evaluation procedure using enthalpy calibration

The heat capacity calibration described above involves three measurements and is therefore time consuming. However, it allows also evaluation in the temperature range where there is only sensible heat stored in the sample, because the calibration is done also with a small signal. The temperature range that is most important is however the melting temperature range with large signals and therefore the calibration should be done for large signals. This can be achieved when using standard materials with a melting effect and using the melting enthalpy for calibration.

The measurement procedure starts with a set of calibration measurements, which has to be repeated usually only every other month. Each calibration measurement is performed on one standard material and is used to determine the sensitivity at one temperature, which is the melting temperature of that standard material, as indicated in fig.3.13.

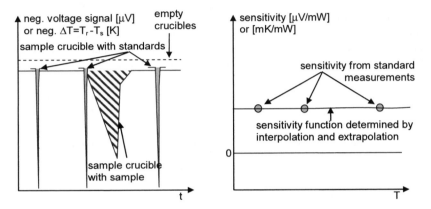

Fig. 3.13. Evaluation of the integrals of the measured signals to determine the sensitivity and finally the enthalpy of the sample.

From eq.3.15 follows, that the integral over the peak area in the measured signal, as indicated in fig.3.13, is proportional to the melting enthalpy ΔH of the standard, also defined the same way:

$$\int_{peakwidth} \Delta T \cdot dt = \int_{peakwidth} \beta \cdot R_{th} \cdot (C_s - C_r) \cdot dt \qquad (3.20)$$

$$= R_{th} \cdot \int_{peakwidth} \frac{dT}{dt} \cdot (C_s - C_r) \cdot dt \, .$$

$$= R_{th} \cdot \Delta H$$

This way of integrating the peak area uses a baseline from the peak onset to the peak end, thereby eliminating the need to make a measurement with empty crucibles. From the integral of the peaks of the standard measurements and the respective literature data for their melting enthalpies, the sensitivity at the melting temperatures of the standard materials can be calculated using eq.3.20. The sensitivity, which is equal to R_{th}, is given in [μV/mW] or [K/W]. Fig.3.13 shows this procedure for a set of different standard materials to cover a temperature range. The sensitivity over a temperature range is then determined by interpolation, or extrapolation if necessary. When a sample is measured, the peak of the sample signal is integrated the same way as it was done with the standards. Then, using the extrapolated sensitivity the melting enthalpy of the sample is calculated using eq.3.20. Because the melting enthalpy as an integral value is only one number, it can be attributed to one temperature. This is usually the onset temperature of the peak, in some cases also the temperature of the peak maximum is used.

While using the enthalpy calibration allows a calibration for large signals, it has some severe disadvantages: it does not take into account sensible heat and it is also not able to give the enthalpy as a function of temperature. These are two restrictions that usually cannot be accepted, because the amount of sensible heat can be significant and because the temperature resolution is crucial in applications with a narrow temperature interval. To estimate how large the difference is between the heat capacity and the enthalpy calibration one can look at the sensitivity. Both ways of calibration should give the same results for the sensitivity as they relate the heat flux to the measured temperature difference. Because the signal in an enthalpy calibration is on the average much higher than in a heat capacity calibration, a systematic difference between sensitivity data calculated from enthalpy calibration and from heat capacity calibration is however observed, as fig.3.14 shows.

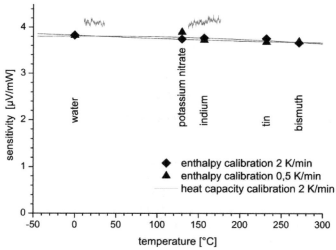

Fig. 3.14. Comparison of sensitivity data calculated from enthalpy calibration and from heat capacity calibration of one DSC (picture: ZAE Bayern)

The typical difference in the sensitivities derived from heat capacity calibration and from enthalpy calibration is in the range of 2 – 10 %. Using the heat capacity calibration can therefore lead to an error of several % in the enthalpy within the melting peak. It would be good to have a higher accuracy, but the total loss of the temperature resolution when using the enthalpy calibration is a much more significant drawback.

3.2.4.2 Differential scanning calorimetry in steps mode

Whatever measurement and evaluation procedure is used in dynamic DSC measurements, a general drawback is that during changing phase the sample is not isothermal. This can lead to a shift of several K of the indicated heat storage capacity, as shown in fig.3.12. A possible solution of this problem is running the DSC in isothermal steps mode. In this case, the temperature of the ambient is changed stepwise in given temperature intervals, as fig.3.15 shows.

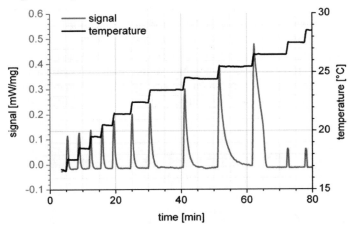

Fig. 3.15. Typical heating ramp and signal during a DSC measurement with the step method (picture: ZAE Bayern)

As with the dynamic mode, the sample follows the temperature change of the ambient and the heat flux to the sample is detected. The difference to a dynamic measurement is that the heating is performed in steps; each step is long enough that the sample comes into thermal equilibrium with the ambient at the end of the step and the signal goes back to zero. At this point, the sample itself is also isothermal. The area below the resulting peak is proportional to the heat absorbed by the sample in the relevant step and can be evaluated using the sensitivity from an enthalpy calibration. After the signal has gone back to zero, indicating that the sample is isothermal, the next step can start.

Measurements with the DSC in steps mode require more programming effort, more time for the measurement itself, and the evaluation is more complex. The big advantage compared to the dynamic mode is that the uncertainty in temperature is given by the height of the steps; reducing the step size therefore results in a higher temperature resolution. Depending on the DSC used, this can be much better than 1 K.

Fig.3.16 shows a comparison of results from measurements with a hf-DSC in dynamic mode, in the steps mode, and with T-history on the same material.

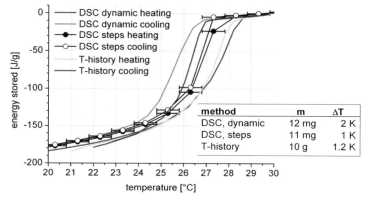

Fig. 3.16. Result from measurements with a hf-DSC in dynamic mode, in the steps mode, and with T-history method on the same material. The symbols for the steps mode mark measured data, the lines the temperature interval (picture: ZAE Bayern).

For the measurements with DSC in dynamic mode, a heating / cooling rate of 0.5 K/min was used. The temperature difference between the enthalpy-temperature relation measured on heating and on cooling is up to 2 K. From these data it is not possible to say if the material has a hysteresis or not, if the difference is caused by the measurement system and parameters, or if it is a mixture of effects between the material and the measurement system. The data from heating and cooling with isothermal steps are practically the same. The uncertainty in the temperature is equal to the step size, here 1 K. These data with their high resolution in temperature clearly show that the investigated material has no significant hysteresis. This means that the deviation observed in the data from dynamic DSC measurements is almost exclusively caused by a too high heating or cooling rate, which means by the selected measurement parameters.

At this point, it is possible to apply what was discussed in section 3.2.3 "Problems in presenting data on PCM" as an exercise to the data shown in fig.3.16. The data from the steps measurement from heating and cooling are practically identical. This means they are the ones closest to thermodynamic equilibrium and thus the ones with the highest accuracy. From reading the data, what is the phase change temperature, the phase change enthalpy, the heat capacity in the solid and the liquid state? There is no melting temperature, but a melting range. There is no well-defined difference between latent and sensible heat, and the heat capacity in the solid cannot be identified as it is not clear where the solid-liquid transition is in a melting range. When giving the *stored heat as a function of temperature in given temperature intervals*, as shown on the right in fig.3.6, these problems are avoided. Tab.3.1 shows the application of this approach to the data from the heating curve in the steps mode in fig.3.16. The density in the calculations was 800 kg/m^3.

Table 3.1. Stored heat as a function of temperature in given temperature intervals, calculated from the data from the heating curve in the steps mode in fig.3.16. The density in the calculations was 800 kg/m^3.

temperature [°C]	temperature interval [°C]	stored heat [J/g]	stored heat [J/ml]	total stored heat [J/g]	total stored heat [J/ml]
20.3	19.8 - 20.8	5	4	-177	-141.6
21.3	20.8 - 21.8	7	5.6	-172	-137.6
22.3	21.8 - 22.8	8	6.4	-165	-132
23.3	22.8 - 23.8	9	7.2	-157	-125.6
24.3	23.8 - 24.8	15	12	-148	-118.4
25.3	24.8 - 25.8	28	22.4	-133	-106.4
26.3	25.8 - 26.8	82	65.6	-105	-84
27.3	26.8 - 27.8	19	15.2	-23	-18.4
28.3	27.8 - 28.8	2	1.6	-4	-3.2
29.3	28.8 - 29.8	2	1.6	-2	-1.6
29.3	29.8 - 30.8	0	0	0	0

3.2.4.3 Differential scanning calorimetry with temperature modulation (m-DSC)

Besides the hf-DSC, which can be operated in dynamic mode and in steps mode, there is another kind of DSC. It is called temperature modulated DSC (m-DSC) and can keep the sample temperature almost constant while the heat capacity or other thermal effects are determined. The measurement is done using a small modulation of the sample temperature, enough to get a small signal to determine the heat flux, but not large enough to influence the investigated thermal effect. A measurement can therefore be done with a sample that is practically isothermal. The disadvantage is that the operation of an m-DSC and the data evaluation and interpretation is significantly more complex than with an hf-DSC. Therefore, it is not described in this introductory text. The interested reader should read special literature on m-DSC, for example Wunderlich et al 1999.

3.2.4.4 T-History method

In section 3.2.1, it was said that it is necessary to have a sufficiently large sample to be representative for the investigated material. A common problem with most DSC is the small size of the sample, usually less than 100 µl. Such small samples are not suitable for inhomogeneous materials as the sample is not representative anymore. Additionally, subcooling observed in a DSC is also not representative for larger samples. For measurements on inhomogeneous materials, a method with much larger sample size is necessary. Such a method is the T-history method.

The T-history method, first proposed by Zhang et al. 1999, is a simple and economic way for the determination of the stored heat as a function of temperature of PCM. Its origin lies in the determination of phase diagrams. When a sample of a material is cooled down and its temperature history (T-history) is recorded, as shown in fig.3.17, any change in its thermophysical properties that leads to a release of heat will result in a change of its temperature history. For example, the release of the latent heat of solidification will ideally result in a sample temperature that remains constant for some time, until the latent heat is released.

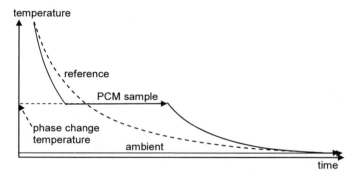

Fig. 3.17. T-history of a sample with a latent heat effect, and a reference with only sensible heat, during cool down at constant ambient temperature.

For a long time this method has been used to find the phase change temperatures of mixtures, data that are necessary to draw phase diagrams. Because the heat exchanged was not measured, it was not a calorimetric method. That changed when Zhang et al. 1999 further developed the idea to be able to get quantitative information on the heat stored. For that, the heat flux during cool down had to be determined. This was done using two identical setups, one with the sample, and one with a reference material (fig.3.18).

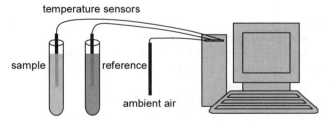

Fig. 3.18. Schematic picture of an experimental T-history set-up after Zhang et al.1999.

The experimental procedure is as follows. The sample, and a reference with known thermal properties, are subject to ambient air. Their temperature history upon cooling down from the same initial temperature to the ambient temperature is recorded, as fig.3.17 shows. A comparison of both curves, assuming identical heat transfer coefficients between sample and ambient, as well as reference and ambient, allows the determination of the heat stored within the sample from the known heat capacity of the reference material. The determination is based on the following calculations:

The heat flux from the sample to the environment, the ambient air, is

$$\dot{Q} = \frac{1}{R_{th}} \cdot \left(T_{sample} - T_{ambient}\right) = -C_{sample} \cdot \frac{d}{dt} T_{sample} . \tag{3.21}$$

To get a quantitative value of the heat flux, the unknown thermal resistance R_{th} has to be determined. For that, the reference with known heat capacity is used. The reference gives

$$\dot{Q} = \frac{1}{R_{th}} \cdot \left(T_{reference} - T_{ambient}\right) = -C_{reference} \cdot \frac{d}{dt} T_{reference} . \tag{3.22}$$

If the sample and reference have the same geometry and are placed in the same environment, R_{th} has the same value for the sample and the reference. From the temperature history and heat capacity of the reference $C_{reference}$, R_{th} can be calculated by eq.3.22. The heat capacity of the sample can then be calculated from the temperature history of the sample and eq.3.21. There are several methods to evaluate the signal of a T-history measurement based on the formulas just described. They differ in what the result is, like one value for the heat capacity of the solid or a temperature dependent value, the way subcooling is treated, and the mathematical procedure. The method described by Zhang et al. 1999 for example uses constant values for the sensible heat capacity in the solid and in the liquid and a melting temperature and melting enthalpy. Marin et al. 2003 have significantly improved the method in two respects. First, they improved the evaluation procedure to be able to calculate the heat capacity as a function of temperature and thus increasing the resolution of the data. Second, they put the samples in a temperature-controlled environment, which significantly enhances the accuracy of the data and further on allows measurements upon heating, and not only on cooling. However, they already pointed out that there were restrictions to the accuracy if the sample is not isothermal. This problem was tackled later at the ZAE Bayern: an insulation around the sample and reference was added thereby reducing the temperature gradient within sample and reference and also reducing variations in the thermal resistance (Günther et al. 2006). Fig.3.19 shows a sketch of the setup. A heat exchanger, connected to a thermostat, is used to control the temperature of the ambient. This allows measurements upon heating and upon cooling. To ensure that the ambient air is isothermal, and that its temperature can be changed fast via the heat exchanger, a fan is used to enforce convection.

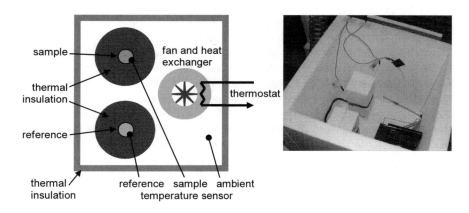

Fig. 3.19. Left: sketch of the T-history setup at the ZAE Bayern designed for heating and cooling experiments in the range of -5 °C to 80 °C. Right: real setup (picture: ZAE Bayern).

Fig.3.20 shows a measurement on a PCM and water as a reference material. For the cooling experiment, the temperature is changed from about 44 °C to 15 °C. For the heating experiment, this is reversed. The temperature step is done within a few minutes; this assures that neither the sample nor the reference has sufficient time to change their temperature. Thus, the temperature step is practically an ideal step.

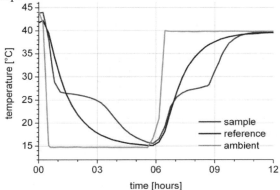

Fig. 3.20. Typical temperature-time curves obtained in a T-history experiment (source: ZAE Bayern).

From fig.3.20, an important advantage of the T-history method becomes clear: in temperature ranges where little heat is stored, usually sensible heat, little time is used during a measurement as the cool down or heat up is rather fast. Only during phase change, where much heat is stored, the change of temperature of the sample is slow. Calorimeters that use fixed heating or cooling rates have a fixed time to scan a temperature range, no matter if that temperature range is important because it has a high storage density, or not. This cannot be changed because the calorimetric signal is proportional to the heating rate and has to be kept constant during a scan. Another advantage of the T-history method is that the signal is not based on the temperature difference between sample and reference, but from the respective temperature differences to the ambient. Therefore, the signal to noise ratio is high even for very slow measurements. In contrast, in a DSC measurement the signal is proportional to the heating rate. Therefore, if a small heating rate is chosen to get an isothermal signal, this will result in a small signal. Because of this advantage, the T-history method is suitable for the acquisition of data close to thermal equilibrium even for large samples. This shows fig.3.16, where results from a T-history measurement are compared to ones from DSC-measurements in the dynamic and in the steps mode. The maximum temperature shift between the enthalpy from the heating and from the cooling measurement determined by the T-history method is 1.2 K; this is more than with the steps mode, but less than with the dynamic mode. The agreement between the total enthalpy difference determined from heating and cooling is good for all three data sets. While the samples for the hf-DSC measurements had a mass of 11 and 12 mg, the sample used in the T-history measurement had a mass of 10 g, almost 1000 times as much. This means that the T-history method can even be used for composite materials with larger inhomogeneities. Further on, while the small sample size in common hf-DSC often increases subcooling, this is rarely observed with the T-History method. Günther et al. 2006 and Hiebler 2007 have done a detailed discussion of this topic.

The comparison of results determined by different methods on the same material or even sample, as just described, gives important information on the influence of sample size and measurement parameters. However, it does not show if the absolute values of the results are correct. For this, a comparison between measured and literature values of standard materials has to be performed. Lazaro et al. 2006a describe such a procedure, useful standard materials, and discuss their results. Information on how the topic of calorimetric measurements is treated in the new standard is available at www.pcm-ral.de.

3.3 Heat storage and heat release of PCM-objects

Generally, it is possible to calculate the performance of a PCM-object, like balls or bags filled with PCM, from material property data and the geometry of the PCM-object. Calculations however can be complicated and time consuming, and sometimes they are not possible when material property data are not available. Even if the calculations are performed, the question arises if the results are correct. Is the effort doing an elaborate calculation worth it? Often it is better to measure the properties of PCM-objects directly, at least to verify the results of a calculation.

Like for the PCM itself, also for PCM-objects the most important property is the storage and release of heat. The heat stored in a PCM-object can basically be determined with the T-history set-up after some modifications. This is straightforward and will not be described here. When it is also necessary to measure the heat release under defined ambient conditions and heat transfer medium, as in a given application, a special setup has to be constructed. The most common cases with air and water as heat transfer medium are now described.

3.3.1 Air and other gases as heat transfer medium

Fig.3.21 shows a setup to test PCM-objects in an air stream, built at the ZAE Bayern. It is designed based on a system developed and described by Zalba 2002. The central component of the setup is a test chamber, through which air of defined temperature and flow rate is blown. From the measured inlet and outlet temperature, and the volume flow rate, the rate of heat transfer (power) can be calculated. By integration of the heat transfer over time, the heat stored in a PCM-object can be calculated. Effects of subcooling on power output can also be observed directly.

86   3 Determination of physical and technical properties

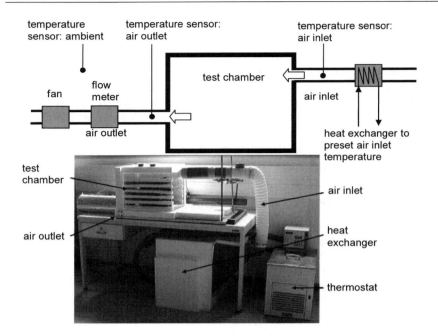

Fig. 3.21. Sketch and picture of the setup at the ZAE Bayern, showing the test chamber with the front insulation removed to see the PCM-objects which are being tested (picture: ZAE Bayern).

Depending on what is put into the test chamber, it is possible to test a single PCM-object in a defined volume flow of air, as shown in fig.3.22 on the left. As the right side shows, it is also possible to test a set of PCM-objects in a special geometry and setup. In that case, the test chamber can act as a small-scale heat storage with air as heat transfer medium.

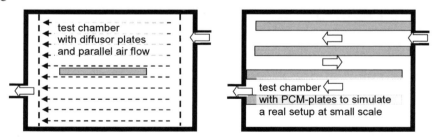

Fig. 3.22. Left: setup to test a single PCM-plate in parallel airflow. Right: setup to test a real storage setup on a small scale.

Fig.3.23 shows a typical data recording during a cooling experiment performed with 4 packages of a salt hydrate. The chamber and the PCM are initially at about 34 °C and then cooled with air of about 16 °C. The top graph shows the temperature recording of the air at the inlet, the outlet, as well as the volume flow of air.

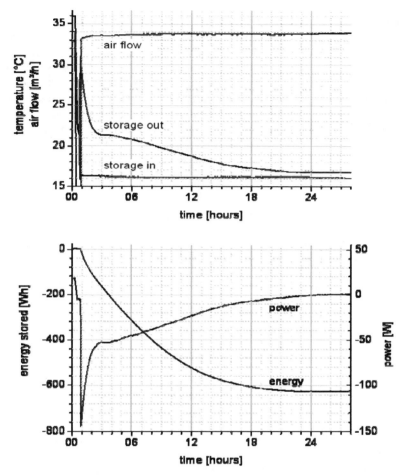

Fig. 3.23. Typical data recording and evaluation for the case of a cooling experiment (pictures: ZAE Bayern).

From the temperature and airflow data, the power output of the test chamber is calculated by

$$P = \dot{Q} = \dot{V}_{air} \cdot c_{p,air} \cdot (T_{out} - T_{in}). \tag{3.23}$$

The graph at the bottom in fig.3.23 shows the calculated power. As pointed out above, instead of making an experiment one could also calculate the power from material data, the geometry of the PCM-objects, and typical heat transfer coefficient. However, is the result of such a calculation or simulation reliable? It is better to test a mathematical model at least by verification with a small-scale experiment. The data also allow to verify the enthalpy data of the used PCM determined by DSC or T-history method. One should not take for granted that data from measurements on small samples will describe accurately the performance of a material when used in large packages. It is always better to compare data from different experiments. For this, the power of the test chamber has to be integrated over time to give the stored heat. The result shows the graph at the bottom in fig.3.23: the total amount of heat stored in the test chamber between about 16 °C and 34 °C was roughly 620 Wh. Taking into account that the test chamber contained 4 packages with 3 kg PCM each, the storage density can be calculated. Fig.3.24 shows the result for the cooling, and additionally also for a heating experiment.

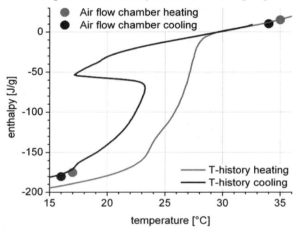

Fig. 3.24. Comparison of the enthalpy difference of the same material determined on large packages of 3 kg each in the test chamber, and of a 15 g sample using the T-history method (picture: ZAE Bayern).

Because in the test chamber the material is not isothermal during the cooling and heating experiment, it is only possible to compare the enthalpy difference between the isothermal state at the beginning and at the end of the measurement (interval mode, section 3.2.1 and 3.2.2). As fig.3.24 shows, the agreement with the data from T-history measurements is very good for the heat stored in the total temperature interval. The data of the stored heat determined with the T-history method on a small sample of 15 g can therefore be used for calculations of larger storages. Günther et al. 2006 give a detailed discussion of this topic.

Fig.3.24 is also useful to discuss the *normalization* of h(T) curves. Here, the data from the T-history measurements have been normalized to 0 J/g at 30 °C, that is a temperature where the sample is liquid but not too far away from the temperature range of the phase change. The reason to choose the temperature this way is that in the liquid state dynamic effects proceed fast. As a consequence, the h(T) functions in the liquid state from heating and from cooling have the same slope and the result of the normalization is not dependent on the exact temperature chosen for normalization. This would be different when a lower temperature is chosen, like 22 °C. At this temperature, the shape of the h(T) curve determined upon cooling strongly depends on dynamic effects, especially on the temperature when nucleation was triggered. The procedure to normalize h(T) curves at a temperature where the sample is liquid, but not too far away from the temperature range of the phase change, has been tested many times and proved to be very helpful. The same procedure was also followed to normalize the h(T) curves in fig.3.16.

3.3.2 Water and other liquids as heat transfer medium

It is possible to construct similar setups for water and other liquids as heat transfer medium. Because the setup is not much different from the one just described for air as heat transfer medium, it will not be describe it here. However, there is a completely different setup to determine the heat storage and heat release of PCM-objects, which cannot be used with gases: the mixing calorimeter.

3.3.2.1 Mixing calorimeter

In a mixing calorimeter, a sample and a liquid reference material of known heat capacity are placed in an adiabatic environment, as shown in fig.3.25. Often, water is used as reference, because its heat capacity is well known and the heat transfer is good. Initially, the sample and the reference liquid have different temperatures, but after a while, thermal equilibrium is reached.

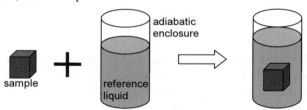

Fig. 3.25. Schematic procedure in a measurement with a mixing calorimeter.

From the law of energy conservation the enthalpy change of the sample between its initial and final temperature can be calculated. It is equal to the heat loss or gain of the reference liquid

$$\Delta H_{sample} = H_{sample}(T_{final}) - H_{sample}(T_{sample,initial})$$
$$= m_{reference} \cdot \int_{T_{reference,initial}}^{T_{final}} c_{p,reference} \cdot dT$$
$$\approx m_{reference} \cdot c_{p,reference} \cdot (T_{reference,initial} - T_{final}) \qquad (3.24)$$

Because the measurement is done between two temperatures, which means in the interval mode, the resolution with respect to temperature is not good. Using different initial temperatures, the enthalpy difference can be determined between different temperature intervals and thus it is possible to establish an enthalpy-temperature relation over a temperature range. The main advantages of this method are that it is not complicated and applicable to PCM-objects with complex shape. Inaba and Tu 1997 describe a setup of a mixing calorimeter and evaluation procedure including heat losses of the vessel and the thermal mass of the vessel.

3.3.2.2 Setup derived from power compensated DSC

Demirel and Paksoy 1993 describe another setup and test procedure to test large PCM-objects in a bath filled with a liquid. The liquid may be water or oil for example. Fig.3.26 shows the setup for the method, which is roughly the transfer of the power compensated DSC method to PCM-objects.

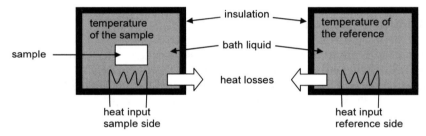

Fig. 3.26. Setup to test PCM-objects derived from power compensated DSC.

For a differential measurement, two identical baths are filled with the same amount of working liquid, which is well stirred. Each bath is heated by an electrical resistance heater, or cooled by copper coils with tap water. Without sample, both baths will show identical behavior upon heating or cooling. For a measurement, one of the baths operates with the sample in the working liquid, the second one with the working liquid alone. In a measurement to scan a temperature range, both baths are heated simultaneously at a constant rate. The heating of the bath with the sample requires a larger heating power, equivalent to the heat stored in the PCM-object. Taking the differential signal between the heating powers of both baths the heat loss, which is the same for both baths, is eliminated from the heating power of the bath with sample. The difference in the heating powers is therefore only due to heat storage in the sample.

To determine the integral of the enthalpy change of the sample between two temperatures another kind of experiment can be performed. First, the sample is preheated to the lower one of the temperatures, while both baths are preheated to the higher temperature. The sample is then put into one of the preheated baths; this will lead to a cool down of the bath. The time integral of the heating power necessary to reach the initial bath temperature again is equal to the enthalpy difference of the sample between the start and end temperature. By taking the difference between the heating powers of both baths, the heat loss of the bath with sample is again taken into account. This means the procedure is similar to the mixing calorimeter; instead of an adiabatic environment, there are heat losses that are determined from the second bath.

These differential measurements provide information on heating and cooling curves including heat transfer characteristics between sample and bath liquid, the thermal behavior of the sample, and enthalpy as a function of temperature or between two temperatures.

3.4 Thermal conductivity of materials

The thermal conductivity λ describes the ability of a material to transport heat while the material itself does not move. The definition of λ is by Fourier's law, which is for 1-dimensional steady state heat conduction written as

$$\dot{Q} = A \cdot \lambda \cdot \frac{dT}{dx} \quad [\lambda] = W/mK \, . \qquad (3.25)$$

The thermal conductivity is a material property, because the temperature gradient defines the temperature change over an infinitesimal length. It is important that the measurement system assures the correct measurement.

Generally, the thermal conductivity λ is higher in the solid than in the liquid state due to better molecular interaction to transport heat. Fig.3.27 shows the schematic temperature dependence of the thermal conductivity and heat capacity for a melting temperature (left) where it is possible to distinguish between values of the solid and the liquid, and for the case of a melting range (right). For the melting range, the transition between solid and liquid is smooth.

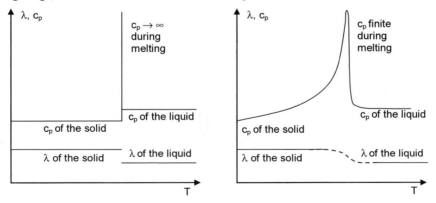

Fig. 3.27. Schematic temperature dependence of the thermal conductivity and heat capacity for a melting temperature (left) and a melting range (right).

The determination of the thermal conductivity can be performed with different methods using stationary, dynamic, or periodic signals.

3.4.1 Stationary methods

All *stationary methods* are based on Fourier's law directly, with a constant temperature difference over some sample length. Then, the gradient in eq.3.25 is replaced by finite differences

$$\dot{Q} = A \cdot \lambda \cdot \frac{\Delta T}{\Delta x}. \tag{3.26}$$

The heat flux and temperature gradient are stationary, which means time independent. Mehling et al. 2000 describe a setup for this method that was used to determine the thermal conductivity of PCM-graphite composites. As fig.3.28 shows, the heat flux is in one dimension between an electrical heater and a cooler that is connected to a thermostat.

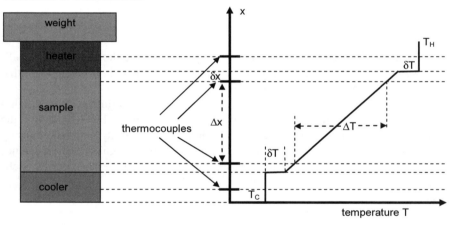

Fig. 3.28. Schematic setup do determine the thermal conductivity with a stationary method (Mehling et al. 2000).

Because of the thermal resistance at the interface between sample and heater, respectively cooler, the temperatures to calculate the temperature gradient have to be measured within the sample. From

$$\dot{Q}_{el} = \dot{Q}_{th} = A \cdot \lambda \cdot \frac{\Delta T}{\Delta x}, \tag{3.27}$$

the thermal conductivity is calculated by

$$\lambda = \frac{\dot{Q}_{th} \cdot \Delta x}{A \cdot \Delta T} = \frac{\dot{Q}_{el} \cdot \Delta x}{A \cdot \Delta T}. \qquad (3.28)$$

Depending on the design, especially how to deal with heat losses, this method can be used in a wide range of thermal conductivities from 0.01 to 10 W/m·K. However, when measuring a PCM, the latent heat can cause a significant delay in getting a stationary signal and linear temperature profile. Too early recording of data can result in large errors in the result. The main disadvantage of this method is however that it is not very suitable for liquids.

3.4.2 Dynamic methods

In contrast to stationary methods, *dynamic methods* are based on time varying temperatures. To describe and evaluate dynamic heat transfer, Fourier's law must be modified. The heat flux in or out of a volume element A·dx is equal to the change per time in the heat stored

$$\frac{dQ}{dt} = \dot{Q} = A \cdot \lambda \cdot \frac{dT}{dx} = A \cdot dx \cdot \rho \cdot c_p \frac{dT}{dt}. \qquad (3.29)$$

Eliminating the area A and separating variables gives

$$\lambda \cdot \frac{dT}{dx} = dx \cdot \rho \cdot c_p \frac{dT}{dt} \qquad (3.30)$$

and finally

$$\frac{\lambda}{\rho \cdot c_p} \cdot \frac{d^2T}{dx^2} = \alpha \cdot \frac{d^2T}{dx^2} = \frac{dT}{dt}. \qquad (3.31)$$

The value $\lambda / (\rho \cdot c_p)$ is called *thermal diffusivity* α. In contrast to stationary methods, dynamic methods do not need stationary conditions and thus are usually faster to get data. Nevertheless, PCM with their large thermal effect in a small temperature range can cause problems.

The most important dynamic method is the *hot-wire method*, because it can easily be applied to solids and liquids in a wide range of thermal conductivities from about 0.03 to 30 W/m·K. That means it fully covers the range of thermal conductivity encountered with PCM and PCM composite materials for the solid and liquid state. Fig.3.29 shows a schematic setup of a hot-wire experiment.

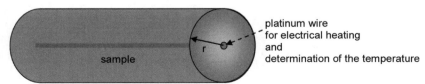

Fig. 3.29. Schematic setup of a hot-wire experiment.

At the beginning of a measurement, the sample is heated to the desired temperature. When the sample is isothermal, the platinum hot wire located at the main axis of the cylindrical sample supplies a constant heat flux dQ/dt electrically. The sample material conducts the heat released by the wire radially away, faster with higher thermal conductivity and slower with higher heat capacity. The conductivity and heat capacity of the sample material therefore influence the temperature rise with time at the main axis. This temperature-time signal is determined using the platinum wire, which is already used as electrical heater, also as temperature probe.

To evaluate the signal it is necessary to solve the dynamic heat transfer equation (eq.3.31) in cylindrical coordinates

$$\alpha \cdot \left(\frac{d^2T}{dr^2} + \frac{1}{r}\frac{dT}{dr} \right) = \frac{dT}{dt}. \tag{3.32}$$

Neglecting boundary effects and the contact resistance wire-sample, the solution is

$$T(t) - T(t=0) = \frac{\dot{Q}}{4 \cdot \pi \cdot \lambda} \cdot \ln\left(\frac{4 \cdot \alpha \cdot t}{r^2 \cdot \ln(0.5772)} \right). \tag{3.33}$$

Taking the difference for different times, all constants in the ln can be eliminated and the result is

$$\Delta T(t) = \frac{\dot{Q}}{4 \cdot \pi \cdot \lambda} \cdot \Delta \ln(t). \tag{3.34}$$

The thermal conductivity λ is therefore calculated from the temperature rise ΔT with time and the heat flux dQ/dt.

Typically, ΔT in a measurement is only a few degrees; therefore, even in liquid samples, convection usually does not start. The melting or solidification of a sample however can cause problems. The reason is that the resulting heat effect is in addition to the heat release of the wire, but not taken into account in the evaluation. It is therefore better to take data points at temperatures out of the melting range.

An example of the application of the hot-wire method to PCM is described by Inaba and Tu 1997. They describe a setup of a hot-wire measurement system and its application to determine the thermal conductivity of shape-stabilized paraffin. More information on how the determination of the thermal conductivity is treated in the new standard is available at www.pcm-ral.de.

3.5 Cycling stability of PCM, PCM-composites, and PCM-objects

The use of a PCM, a PCM-composite material, or a PCM-object in any application is based on the assumption that it will perform as expected for the lifetime of the application. During the lifetime, many heating and cooling cycles with repeated melting and solidification have to be performed. The reproducibility of the phase change, also called cycling stability, is tested by cycling tests. In a cycling test a PCM, PCM-composite, or PCM-object is subject to many heating and cooling cycles and its performance is monitored continuously or in fixed intervals. The word "performance" can have many meanings regarding cycling stability. Performance refers to all important physical properties, for example enthalpy as a function of temperature and thermal conductivity. A PCM for example which shows phase separation will show a reduction of the melting enthalpy after repeated cycling. If the reduction is too high, the PCM is not stable during cycling. For PCM-objects, cycling stability can refer to additional aspects. For example for encapsulated PCM, the encapsulation can be tested for leakage after repeated melting and solidification.

3.5.1 Cycling stability with respect to the stored heat

To test the cycling stability with respect to the stored heat, it is necessary to determine the stored heat after a number of cycles. To make sure that no other effects from the environment cause any change, an encapsulation that ensures the constant composition of the PCM must be provided.

The easiest way to test a PCM or PCM-composite is to determine the enthalpy in a calorimeter and to do the cycling in the calorimeter also. A very common example is to do this in a DSC by repeating the measurement of the enthalpy by heating and cooling many times. In this case, the crucibles act as encapsulation. If the T-History method is used for the determination of the enthalpy, the samples are already in a kind of test tube. The cycling of the test tubes can be done in a separate water bath or a thermostat with automatic heating and cooling. One can also use a hot bath for heating the sample, and after melting, put the sample in a second, cold bath for cooling and crystallization. The advantage is that time and energy is saved since no bath has to be heated or cooled. It is only necessary to compensate heat losses or gains. In contrast to the cycling in the DSC, the cycling in the water bath only ensures that the sample is heated and cooled. If the sample was really melted and solidified has to be checked by other means. In the DSC, the signal indicates if the phase change took place or not. To test the cycling stability of PCM objects it is necessary to place the object in a heated or cooled environment. This can be a bath again or a climate chamber.

3.5.2 Cycling stability with respect to heat transfer

If the thermal conductivity of a PCM or PCM composite is an important physical property, for example as in the PCM-graphite composites, the stability of the thermal conductivity upon cycling has to be tested.

One way to test this is to cycle the sample in a thermostat or a climate chamber and then take it out every other cycle to measure the thermal conductivity. This approach is accurate, however can be cost intensive because of the frequent handling of the sample.

An alternative method is to do the testing and cycling in the same apparatus, similar to the cycling and testing for enthalpy in the DSC. A possible setup to do the cycling shows fig.3.30. It consists of one or several sample containers, heated and cooled successively by hot or cold water from two thermostats.

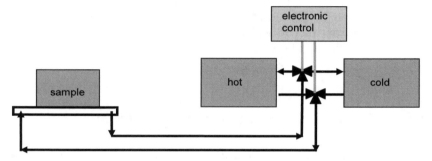

Fig. 3.30. Setup to test cycling stability with one sample container, which is heated or cooled successively by hot or cold water from two thermostats.

Fig.3.31 shows a real installation of such a setup. The left side shows the whole installation with the thermostats, the samples, and the electronic control. The right side shows a close up of the samples.

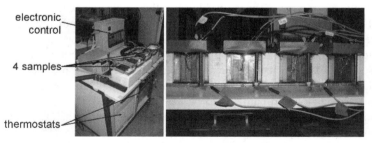

Fig. 3.31. Experimental setup for automatic heating and cooling of four samples and recording of the temperature over time (pictures: ZAE Bayern)

To test the stability of the thermal conductivity it is possible to take out the samples after a number of cycles and measure the thermal conductivity. However, testing cycling stability does not necessarily require a determination of an absolute value; detecting if a property has changed or not can be sufficient. In this case, to test the stability of the thermal conductivity of the samples, each sample is equipped with thermocouples at different distances from the heat exchanger. If the heating and cooling is done in a reproducible way, any change of the thermal conductivity or the phase change enthalpy of a sample during cycling will be visible in a comparison of the recorded temperature-time curves. If after a number of cycles a change is observed, one can decide to take a sample out and to determine the value of the thermal conductivity.

Information on how the topic of cycling stability is treated in the new standard is available at www.pcm-ral.de.

3.6 Compatibility of PCM with other materials

Another important topic in PCM technology is the compatibility of a PCM with other materials. This is with respect to three aspects. The first one is the tightness of an encapsulation. Migration of the PCM or components of it through the wall of the encapsulation is possible and can lead to a loss in storage capacity as the PCM composition or its amount is changed. An example is the loss of water of a salt hydrate when the encapsulation material allows water molecules to diffuse to the outside. The second aspect is the stability of the encapsulation in contact with the PCM. Here, the negative change of plastic properties when in contact with organic PCM and the corrosion of metals in contact with inorganic PCM are common. These are also the most common risks with regard to the third aspect, the compatibility of the PCM with materials outside the encapsulation in case of leakage.

To test compatibility between materials, in many cases standardized procedures are available. It is often possible to apply these standards also to PCM, but keeping in mind that some things might be different when testing PCM. For example, common standards test the influence of one material on some property of another material, for example the influence of a corrosive liquid on the thickness of a metal sample. When the liquid is a PCM, it might be possible that corrosion products also have an influence on the PCM. Further on, in a real application the PCM will change phase and thus can have a mechanical effect on a surface. In combination with electrochemical corrosion, this can lead to a different result than just with electrochemical corrosion.

3.6.1 Corrosion of metals

Many inorganic PCM such as eutectic water-salt solutions, salt hydrates, and salts have the problem of causing corrosion when in contact with metals. Often people think this is a general problem, but this is not correct, as shown later. Some data on the compatibility of these PCM with metals can be found in the literature, but often the composition of the PCM is not well documented or it has been changed. For example, often tests of salt hydrates have not been performed with the correct water concentration but with a solution containing additional water. If the correct information is not available, a compatibility test is necessary. Such a test is carried out with only the metal pieces immersed in the PCM, or in combination with other materials. For example, when a PCM-graphite composite will be used in an application, graphite pieces have to be immersed in the PCM in addition to the metal.

The following section gives an example of such a test procedure based on typical corrosion tests. It has been used to test the combination of salt solutions and of salt hydrates in combination with metals. First, the metal pieces and the melted PCM are placed in closed test tubes. The test tubes are then immersed in a water bath, as shown in fig.3.32, to keep the temperature constantly at about 20 K above the melting temperature of the PCM.

Fig. 3.32. Compatibility test for inorganic PCM in combination with metals, and additionally in contact with a piece of graphite (pictures: ZAE Bayern).

Five samples of each combination were tested and the results evaluated are the average of them. If the results differ strongly, a more detailed investigation might be necessary. To get information on the time dependence of corrosion, metal samples should be removed from the test tubes after different times; for example one or two samples after 1 month, another one or two after 2 months, and maybe a third set after 6 months. After taking the sample out of the water bath, they should be evaluated with the following procedure:

- Testing the pH of the PCM
- Evaluation of changes in the appearance and characteristics of the PCM to identify qualitatively the precipitate formed (if there is any precipitate)
- Cleaning of the metal pieces with tap water and visual evaluation of their change in appearance (fig.3.32, bottom)
- Polishing the metal pieces with 150 grain abrasive paper
- Gravimetric analysis before and following the corrosion tests provides mass loss, Δm, with respect to the initial mass

$$\Delta m = m(t_0) - m(t). \tag{3.35}$$

- Measurements before and following the tests provide reduction in sample thickness, and length (mm)
- Calculation of the corrosion rate, defined as mass loss per square meter and day

$$CR = \frac{m(t_0) - m(t)}{(t_0 - t) \cdot A}. \tag{3.36}$$

The results can be presented in a graph to visualize the change of the corrosion rate with time. This has for example been done in Cabeza et al. 2002 to clarify if corrosion gets worse with time or not. A more compact way of presentation and evaluation is the use of a guide for corrosion weight loss used in the industry (tab.3.2).

Table 3.2. Guide for corrosion weight loss used in the industry (source: product information sheet, Carborundum Corporation, USA).

mg/cm²yr	mm/yr	Recommendation
> 1000	2	Completely destroyed within days
100 to 999	0.2 – 1.99	Not recommended for service greater than a month
50 to 99	0.1 – 0.19	Not recommended for service greater than one year
10 to 49	0.02 – 0.09	Caution recommended, based on the specific application
0.3 to 9.9	-	Recommended for long term service
< 0.2	-	Recommended for long term service; no corrosion, other than as a result of surface cleaning, was evidenced

Cabeza et al. 2001a, Cabeza et al. 2001b, Cabeza et al. 2001c, Cabeza et al. 2002, and Cabeza et al. 2005 have tested different PCM with this methodology. These publications also include tests with graphite added to the PCM – metal sample. Here a few results are given. Tab.3.3 lists the PCM tested. They are from the groups of salt hydrates and eutectic water-salt solutions.

Table 3.3. Properties of the PCM investigated.

Salt hydrate	Chemical formula	Melting temperature	pH of liquid PCM
Sodium acetate trihydrate	$NaOAc \cdot 3H_2O$	58 °C	10.5
Zinc nitrate hexahydrate	$Zn(NO_3)_2 \cdot 6H_2O$	36 °C	3
Calcium chloride hexahydrate	$CaCl_2 \cdot 6H_2O$	32 °C	6
Potassium hydrogen carbonate (16.5 wt.%) / H_2O	$KHCO_3$	-6 °C	9
Potassium chloride (6 wt.%) / H_2O	KCl	-10 °C	6

Tab.3.4 shows the results obtained for the different PCM-metal combinations. The data show the full range of possible results from recommended to not recommended at all. For example $Zn(NO_3)_2 \cdot 6H_2O$ showed very little effect on stainless steel, but significantly corroded copper. This shows that the compatibility of salt solutions and salt hydrates with metals is not in general a problem, as it is often assumed. Actually many combinations are compatible and can be recommended. On the other hand, often people assume that organic PCM and metals are no critical combination; however, this is not completely correct. For example, fatty acids do have some corrosive effect on metals even though it is not too strong.

Table 3.4. Suitability of the PCM - metal pairs studied.

PCM / metal	brass	copper	stainless steel	steel	aluminum
Sodium acetate trihydrate	15 to 4 mg/cm²yr Caution recommended	33 mg/cm²yr Caution recommended	< 0.2 mg/cm²yr Recommended	< 0.2 mg/cm²yr Recommended	< 0.2 mg/cm²yr Recommended
Zinc nitrate hexahydrate	Not recommended	100 to 300 mg/cm²yr Not recommended	< 0.2 mg/cm²yr Recommended	Not recommended	35 to 100 mg/cm²yr Not recommended
Calcium chloride hexahydrate	0.4 to 2.5 mg/cm²yr Recommended	0.4 to 5 mg/cm²yr Recommended	0 to 0.4 mg/cm²yr Recommended	26 to 37 mg/cm²yr Caution recommended	4 to 11 mg/cm²yr Caution recommended (gas production)
Potassium hydrogen carbonate (16.5 wt.%) / H_2O	15 mg/cm²yr Caution recommended	10 to 13 mg/cm²yr Caution recommended	0.3 to 2 mg/cm²yr Recommended	< 0.2 mg/cm²yr Recommended	< 0.2 mg/cm²yr Recommended
Potassium chloride (6 wt.%) / H_2O	24 mg/cm²yr Caution recommended	11 to 24 mg/cm²yr Caution recommended	2 to 6 mg/cm²yr Recommended	26 to 32 mg/cm²yr Caution recommended	6 to 14 mg/cm²yr Caution recommended

3.6.2 Migration of components in plastics

Another problem of materials compatibility exists with plastics. Due to the microscopic structure of some plastics, it is possible that organic PCM or water molecules migrate through plastic encapsulations. While some general information on the diffusion of water and organic liquids through plastics can be found in the literature, data on PCM is very rare. Lázaro et al. 2005 have described a test procedure, which is based on the standard ISO 175:1999, and used it to test the compatibility of several PCM with different plastics. The following text outlines the content of that publication. The methodology used consists of the following steps:

- Melting of the PCM and then stirring it to get a homogeneous PCM sample
- Introduction of about 30 ml of PCM into a plastic bottle made of the plastic to be tested. Fig.3.33 shows the bottles containing PCM used in the tests

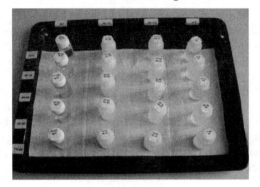

Fig. 3.33. Compatibility test for plastic-inorganic PCM and plastic-organic PCM combinations (picture: Univ. Zaragoza).

- Cycling the bottles with the PCM in a furnace or any other similar device, to have repeated melting and solidification
- Frequent visual inspection of the bottles, in order to see if plastic deformations or other phenomena take place
- Gravimetric analysis of the bottles containing the PCM before and following the compatibility test provides mass variation. The mass loss is calculated as defined in eq.3.35. The measurements should be made when the bottles are at room temperature in order to diminish deviations due to variations of temperature. Once the bottles are at room temperature, an adsorbent paper can be used

to remove the PCM outside the bottles and the dust particles deposited on the walls. Then the total mass of each bottle with each PCM is measured with a precision balance.

Lázaro et al. 2005 used this methodology to test the compatibility of the four organic PCMs molecular alloy C16-C18, RT20, RT26, RT25, and the inorganic PCM TH24, each in combination with the four plastics LDPE, HDPE, PET, and PP. During the 10 months test period the samples have been melted and solidified about 150 times, and visual inspection and gravimetric analysis were made. The migration of organic PCM and of moisture, both have been detected. The mass changes were up to several wt.% of the PCM mass. LDPE bottles had the highest mass changes and big deformations with all the PCM tested. Therefore, LDPE should not be chosen as encapsulate material with any of the PCMs tested. PET showed the best results for all organic PCM, and for TH24, HDPE showed the lowest mass changes. The tests are also described in Lazaro et al. 2006b.

The method of using bottles made of the material which is investigated is straight forward for testing the migration of PCM or PCM components to the outside or of substances from the outside to the inside of an encapsulation, because only the change in mass has to be measured. When it is necessary to test more or different properties of a material for compatibility, it is difficult to cut out a piece of a bottle for any of the tests. Further on, for each test a whole bottle is lost. It is then better to put regular shaped pieces of the plastic that is investigated into a vessel containing the PCM. The advantage is that the pieces can be tested for mass change and further on change in mechanical properties, optical properties, or heat capacity. Otherwise, the same procedure like repeated melting and solidification can be used as described above.

3.7 References

[Cabeza et al. 2001a] Cabeza L.F., Illa J., Roca J., Badia F., Mehling H., Hiebler S., Ziegler F.: Immersion corrosion tests on metal-salt hydrate pairs used for latent heat storage in the 32 to 36 °C temperature range. Materials and Corrosion **52**, 140 – 146 (2001)

[Cabeza et al. 2001b] Cabeza L.F., Badia F., Illa J., Roca J., Mehling H., Ziegler F.: Corrosion experiments on salt hydrates used as phase change in cold storage. Presented at Planning Workshop IEA ECES Annex 17 "Advanced thermal energy storage techniques – Feasibility studies and demonstration projects", Lleida, Spain, 5.-6. April 2001. www.fskab.com/annex17

[Cabeza et al. 2001c] Cabeza L.F, Illa J., Roca J., Badia F., Mehling H., Hiebler S., Ziegler F.: Middle term Immersion corrosion tests on metal-salt hydrate pairs used for latent heat storage in the 32 to 36 °C temperature range. Materials and Corrosion **52**, 748-754 (2001)

[Cabeza et al. 2002] Cabeza L.F, Roca J., Noques M., Mehling H., Hiebler S.: Immersion corrosion tests on metal-salt hydrate pairs used for latent heat storage in the 48 to 58°C temperature range. Materials and Corrosion **53**, 902 – 907 (2002)

[Cabeza et al. 2005] Cabeza L.F., Roca J., Nogues M., Mehling H., Hiebler S.: Long term immersion corrosion tests on metal-PCM pairs used for latent heat storage in the 24 to 29 °C temperature range. Materials and Corrosion, vol.56, No. 1 (2005)

[Cammenga et al. 1993] Cammenga H.K., Eysel W., Gmelin E., Hemminger H., Höhne G., Sarge S.: The temperature calibration of scanning calorimeters. Part 2: Calibration substances. Thermochimica Acta **219**, 333-342 (1993)

[Demirel and Paksoy 1993] Demirel Y., Paksoy H.Ö.: Thermal analysis of heat storage materials; Thermochimica acta **213**, 211-221 (1993)

[Gmelin and Sarge 1995] Gmelin E., Sarge S.M.: Calibration of differential scanning calorimeters; Pure & Appl. Chem., Vol. 67, No. 11, 1789-1800 (1995)

[Gmelin and Sarge 2000] Gmelin E., Sarge S.M.: Temperature, heat and heat flow rate calibration of differential scanning calorimeters; Thermochimica Acta **347**, 9-13 (2000)

[Günther et al. 2006] E. Günther, S. Hiebler, H. Mehling: Determination of the heat storage capacity of PCM and PCM-objects as a function of temperature. Proc. of ECOSTOCK, 10th International Conference on Thermal Energy Storage, Stockton, USA, 2006

[Hemminger and Cammenga 1989] Hemminger W.F., Cammenga H.K.: Methoden der der Thermischen Analyse. Springer-Verlag, Berlin Heidelberg New York (1989) ISBN 3-540-15049-8

[Hiebler 2007] Hiebler S.: Kalorimetrische Methoden zur Bestimmung der Enthalpie von Latentwärmespeichermaterialien während des Phasenübergangs. PhD Thesis, TU Munich (2007)

[Höhne 1990] Höhne G.W.H.: Fachbeitrag Die Temperaturkalibrierung dynamischer Kalorimeter. Empfehlung des Arbeitskreises "Kalibrierung dynamischer Kalorimeter" der Gesellschaft für Thermische Analyse e.V. (GEFTA), PTBMitteilungen, 100, 1, (1990)

[Inaba and Tu 1997] Inaba H., Tu P.: Evaluation of thermophysical characteristics on shape-stabilized paraffin as a solid-liquid phase change material. Heat and Mass Transfer **32**, 307-312 (1997)

[Lázaro et al. 2005] Lázaro A., Zalba B., Marín J.-M^a, Cabeza L. F.: Phase Change Materials and plastics compatibility. Presented at 8th Workshop IEA ECES Annex 17 "Advanced thermal energy storage techniques – Feasibility studies and demonstration projects", Kizkalesi, Turkey, 18-20 April 2005. www.fskab.com/annex17

[Lázaro et al. 2006a] Lázaro A., Günther E., Mehling H., Hiebler S., Marin J.M., Zalba B.: Verification of a T-history installation to measure enthalpy versus temperature curves of phase change materials. Meas. Sci. Technol. **17**, 2168–2174 (2006)

[Lázaro et al. 2006b] Lázaro A., Zalba B., Bobi M., Castellon C., Cabeza L.F.: Experimental study on phase change materials and plastics compatibility. Environmental and energy engineering, vol. 52, no. 2, 804-808 (2006)

[Marin et al. 2003] Marin J.M., Zalba B., Cabeza L., Mehling H.: Determination of the enthalpy-temperature curves of phase change materials with the T-history method – improvement to temperature dependent properties. Meas. Sci. Technol. **14**, 184 – 189 (2003)

[Mehling et al. 2000] Mehling H., Hiebler S., Ziegler F.: Latent heat storage using a PCM-graphite composite material. Proc. of TERRASTOCK 2000, Stuttgart, 2000.

[Mehling et al. 2006] Mehling H., Ebert H.-P., Schossig P.: Development of standards for materials testing and quality control. Proc. of 7th IIR Conference on Phase Change Materials and Slurries for Refrigeration and Air Conditioning, Dinan, France, September 2006

[Schubnell 2000] Schubnell M.: Temperature and heat flow calibration of a DSC-Instrument in the temperature range between -100 and 160°C. J. Thermal Analysis and Calorimetry **61**, 91-98 (2000)

[Schneker and Stäger 1997] Schneker B., Stäger F.: Influence of the thermal conductivity on the Cp-determination by dynamic methods. Thermocimica Acta 304/305, 219-228 (1997)

[Speyer 1994] Speyer R. F.: Thermal Analysis of Materials. Marcel Dekker Inc., New York (1994)

[Wunderlich et al. 1999] Wunderlich B., Boller A., Okazaki I., Ishikiriyama K., Chen W., Pyda M., Pak J., Moon I., Androsch R.:Temperature-modulated differential scanning calorimetry of reversible and irreversible first-order transitions. Thermochimica Acta **330**, 21-38 (1999)

[Zalba 2002] Zalba B.: Thermal energy storage with phase change - Experimental procedure. Ph D. PhD Thesis, University of Zaragoza (2002)

[Zhang et al. 1999] Zhang Y., Jiang Y., Jiang Y.: A simple method, the T-history method, of determining the heat of fusion, specific heat and thermal conductivity of phase change materials. Measurement Science and Technology **10**, 201-205 (1999)

4 Heat transfer basics

Applications of PCM cover many diverse fields, but all have in common that the most important selection criterion for a PCM is the phase change temperature. Only an appropriate selection ensures repeated melting and solidification. Connected to the melting and solidification process is the transfer of heat. Only the transfer of heat allows the storage of heat in a PCM, or the stabilization of the temperature of something else by a PCM. Depending on the application, the heat flux ranges from several kW for space heating with water, down to the order of several W for temperature control in transport boxes. In this chapter, the basics of heat transfer in PCM are discussed; it is a revision and extension of the description in Mehling et al. 2007. It will be shown how to calculate the heat flux and the time to complete the phase change in simple problems. In chapter 5, this knowledge will be used as a basis to look at the design of complete storages and discuss general design strategies.

To discuss the basics of heat transfer in PCM, it is straightforward to start with the very basic component of any latent heat storage: a volume element of the latent heat storage material that exchanges heat with the surrounding across its surface.

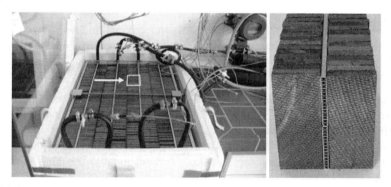

Fig. 4.1. Heat storage model with heat transfer between a volume element of a latent heat storage material, indicated by the arrow, and a plate heat exchanger (pictures: ZAE Bayern).

Fig.4.1 shows a heat storage model with heat transfer between a volume element of a latent heat storage material, indicated by the arrow, and a plate heat exchanger. Generally, the heat exchange across the surface of the volume element can be to a heat exchanger, an insulation layer, or other volume elements of the storage material. Further on, the volume element can be spherical, cylindrical, plate-like, or any other geometry. The basic heat transfer effects are now explained using a 1-dimensional model first, which resembles the geometry in fig.4.1.

4.1 Analytical models

4.1.1 1-dimensional semi-infinite PCM layer

Fig.4.2 shows the heat transfer in a PCM in a 1-dimensional semi-infinite layer upon cooling. The layer is cooled at the surface on the left and extends to the right to infinity.

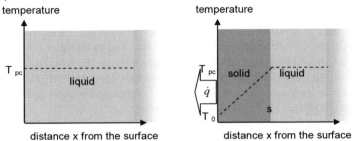

Fig. 4.2. Cooling of a semi-infinite PCM layer. Left: initial situation at t = 0, right: situation at a later time.

Besides the geometrical restriction of a 1-dimensional and semi-infinite PCM layer, and no change in volume due to the solid-liquid phase change, it is necessary to apply some thermal restrictions to arrive at an analytical solution for the heat transfer. These restrictions are:

1. Heat is stored only as latent heat. The sensible heat stored is negligible compared to the phase change enthalpy and therefore only latent heat at the phase change temperature is considered in heat transfer.
2. Heat transfer is only by conduction, there is no convection. Then, the temperature profiles are linear and the heat flux is proportional to the temperature gradient.
3. At the beginning, at t = 0, the PCM is liquid and its temperature is the phase change temperature T_{pc} (fig.4.2, left).
4. The temperature at x = 0 is changed to T_0 and kept constant at that level for any later time (fig.4.2, right).

Fig.4.2 on the right shows the temperature profile after some time; the solidification front has preceded a distance s away from the surface. Fig.4.3 shows the case of heating.

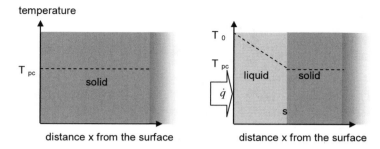

Fig. 4.3. Heating of a semi-infinite PCM layer; initial situation at the left and later at the right.

This simple heat transfer problem under the restrictions listed above is called *Stefan problem*. The mathematical solution (Stefan 1891) is straightforward and the result is very useful in many cases. Therefore, the solution is derived and discussed here:

From neglecting the sensible heat everywhere (restriction 1)

$$c_p \cdot (T_{pc} - T_0) << \Delta_{pc} h \tag{4.1}$$

follows, that the amount of heat released dQ when the phase front moves a distance ds is:

$$dQ(t) = \Delta_{pc} h \cdot A \cdot ds(t). \tag{4.2}$$

The heat released when the phase front is moving is equal to the heat that leaves at the surface, because according to eq.4.1 no heat is stored in between. The heat flux density (heat flux per area) at the surface dq/dt is then calculated from eq.4.2 by taking the time derivative and dividing it by the area

$$\frac{dq(t)}{dt} = \dot{q}(t) = \Delta_{pc} h \cdot \frac{ds}{dt}. \tag{4.3}$$

On the other hand, because the sensible heat is negligible and heat transfer is only by conduction, the temperature change from the location of the phase front at distance s to the surface is linear (fig.4.2 and fig.4.3). Then, the heat flux density at the surface as a function of the distance of the phase front s is

$$\dot{q}(s) = \lambda \cdot \frac{T_{pc} - T_0}{s}. \tag{4.4}$$

Setting eq.4.3 and eq.4.4 for the heat flux density equal gives

$$\lambda \cdot \frac{T_{pc} - T_0}{s} = \Delta_{pc} h \cdot \frac{ds}{dt}. \tag{4.5}$$

Separating the variables in space s and time t to different sides and integrating from t′=0 to t′=t

$$\int_{t'=0}^{t'=t} \frac{\lambda \cdot (T_{pc} - T_0)}{\Delta_{pc} h} \cdot dt' = \int_{s(t'=0)}^{s(t'=t)} s \cdot ds \qquad (4.6)$$

gives

$$\frac{\lambda \cdot (T_{pc} - T_0)}{\Delta_{pc} h} \cdot t = \frac{1}{2} \cdot s(t)^2 . \qquad (4.7)$$

The time for the phase front to proceed a distance s away from the surface is therefore

$$t = \frac{1}{2} \cdot \frac{\Delta_{pc} h}{\lambda \cdot (T_{pc} - T_0)} \cdot s(t)^2 . \qquad (4.8)$$

The location of the phase boundary s as a function of time t is then

$$s(t) = \sqrt{2 \cdot \frac{\lambda \cdot (T_{pc} - T_0)}{\Delta_{pc} h} \cdot t} . \qquad (4.9)$$

Substituting this into eq.4.4, the heat flux density as a function of time t is

$$\dot{q}(t) = \sqrt{\frac{(T_{pc} - T_0) \cdot \Delta_{pc} h \cdot \lambda}{2 \cdot t}} . \qquad (4.10)$$

Eq.4.9 and eq.4.10 show the basic influence of T_{pc}, T_0, λ, $\Delta_{pc} h$, and t on the variation of the heat flux density and location of the phase front with time. Often, people are surprised that $\Delta_{pc} h$ is part of these solutions, but this can be understood easily. The larger $\Delta_{pc} h$, the longer it takes until the phase front has moved a certain distance away from the surface at a given heat flux density. As this distance enters into the heat flux density via the temperature gradient, $\Delta_{pc} h$ also has an influence on the heat flux density.

4.1.2 1-dimensional semi-infinite PCM layer with boundary effects

Until now, it was assumed that the temperature change is directly at the surface of the storage material. In a real design with a heat exchanger, as shown in fig.4.1, this is not the case. In a real case, the boundary consists of a heat exchanger with a heat transfer fluid, as shown in fig.4.4.

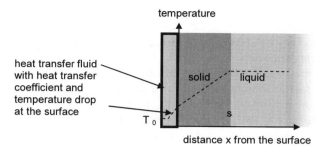

Fig. 4.4. Cooling of a semi-infinite PCM layer with boundary effects.

The difference to the case discussed before is that additional thermal resistances now exist in the heat transfer fluid and the heat exchanger wall, and that the temperature change from T_{pc} to T_0 is not just across the storage material.

The effect of that series of thermal resistances between T_0 and T_{pc} makes a modification of eq.4.4 necessary. Using the thermal resistance of the heat exchanger wall $d_{wall} / \lambda_{wall}$ and of the fluid $1 / \alpha_{fluid}$, the heat flux density into the fluid is now

$$\dot{q}(t) = \frac{1}{\frac{s(t)}{\lambda} + \frac{d_{wall}}{\lambda_{wall}} + \frac{1}{\alpha_{fluid}}} \cdot (T_{pc} - T_0). \tag{4.11}$$

Defining the overall heat transfer coefficient k by

$$\frac{1}{k} = \frac{d_{wall}}{\lambda_{wall}} + \frac{1}{\alpha_{fluid}} \tag{4.12}$$

the solution is (Baehr and Stephan 1994)

$$t = \frac{\Delta_{pc} h \cdot s^2}{2 \cdot \lambda \cdot (T_{pc} - T_0)} \cdot \left(1 + 2 \frac{\lambda}{k \cdot s}\right). \tag{4.13}$$

In the ideal case the heat transfer resistance at the surface, which consists of the thermal resistances of the fluid and the wall, vanishes. As a consequence $k \to \infty$ (eq.4.12) and eq.4.13 becomes equal to eq.4.8 again. The solution in eq.4.13 including boundary effects can also be solved for the location of the phase front s as a function of time (Mehling et al. 1999)

$$s(t) = \frac{1}{2 \cdot \Delta_{pc} h \cdot k} \cdot$$

$$\left[-2 \cdot \Delta_{pc} h \cdot \lambda + 2 \cdot \sqrt{\Delta_{pc} h \cdot \lambda \cdot \left(\Delta_{pc} h \cdot \lambda + 2 \cdot k^2 \cdot t \cdot (T_{pc} - T_0)\right)}\right] \tag{4.14}$$

Despite the significant restrictions discussed at the beginning of section 4.1.1 the analytic solutions given in eq.4.8, eq.4.9, eq.4.10, eq.4.13, and eq.4.14 can be quite valuable. They can be used as a first estimate, especially to study the influence of different parameters.

As an example, a case study of the influence of the heat transfer coefficient on the time necessary to solidify a certain layer thickness of PCM in a 1-dimensional geometry is now performed. Before doing the necessary time dependent calculations, it is necessary to look at typical values of the heat transfer coefficients. As tab.4.1 shows, the heat transfer coefficient can vary within a wide range.

Table 4.1. Heat transfer coefficients for layers of different materials and thicknesses and for heat transfer fluids (heat transfer fluid data on α from Çengel 1998).

layer	thickness d	thermal conductivity λ [W/mK]	heat transfer coefficient λ / d [W/m^{2}K]
PCM	0.001 m = 1 mm	0.5	500
	0.001 m = 1 mm	5	5000
	0.01 m = 1 cm	0.5	50
	0.01 m = 1 cm	5	500
	0.1 m = 10 cm	0.5	5
	0.1 m = 10 cm	5	50
insulating material	0.01 m = 1 cm	0.050	5
	0.1 m = 10 cm	0.050	0.5
gap filled with gas	0.01 mm	~ 0.01 - 0.03	~ 1000 ... 3000
	0.01 m = 10 mm	~ 0.01 - 0.03	~ 1 ... 3
gap filled with liquid	0.01 mm	> 0.2	> 20000
	0.01 m = 10 mm	> 0.2	> 20
heat transfer fluid	comment		heat transfer coefficient α [W/m^{2}K]
gas	free convection		~2 – 25
	forced convection		~25 – 250
liquid	free convection		~10 – 1000
	forced convection		~50 – 20000
vapour	boiling and condensation		~ 2500 – 100000

A comparison of the heat transfer coefficient of layers of PCM with different thickness to heat transfer coefficients at surfaces shows that for a layer thickness of 1 cm and a thermal conductivity of 5 W/m·K the heat transfer coefficient within the PCM is much larger than to a gas as heat transfer fluid by free convection. In that case, the thermal resistance in the PCM can be neglected. The low heat transfer coefficient of the gas on the surface restricts the overall heat transfer. For a layer thickness of 10 cm and a thermal conductivity of 0.5 W/m·K, the heat transfer coefficient within the PCM is significant. If the heat transfer medium is a liquid, even a PCM layer with thickness 1 cm and thermal conductivity of 0.5 W/m·K can become a significant thermal resistance. PCM layers with larger thickness or lower thermal conductivity will dominate the overall heat transfer resistance.

Now, it is possible to study the time dependent influence of the different heat transfer coefficients, where the time dependence enters via the varying thermal resistance of the moving phase change front. The thermal conductivity λ is varied between 0.1 W/mK, 0.5 W/mK, 5 W/mK and 25 W/mK. These values are typical for organic and inorganic PCM and the two different PCM-graphite composites. The phase change enthalpy and temperature drop will be constant as $\Delta_{pc}h = 200$ MJ/m^3 and $T_{pc} - T_0 = 5$ K. Further on, the heat transfer resistance of the wall of the containment is neglected for simplicity; then $k = \alpha_{fluid}$ in eq.4.12. For the study, α is varied between 2500 W/m^{2}K, 250 W/m^{2}K, 25 W/m^{2}K and 5 W/m^{2}K; these are typical values in many applications. Under the different conditions, the time necessary to solidify a certain layer thickness of PCM in a 1-dimensional geometry is as depicted in fig.4.5 to fig.4.9.

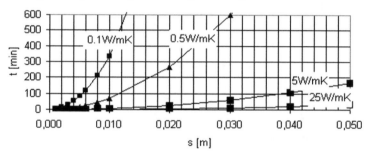

Fig. 4.5. Ideal case, $\alpha \rightarrow \infty$.

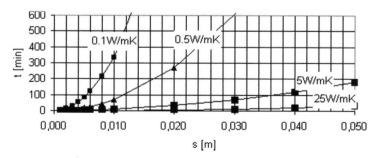

Fig. 4.6. $\alpha = 2500$ W/m^{2}K, typical for forced convection of a liquid.

112   4 Heat transfer basics

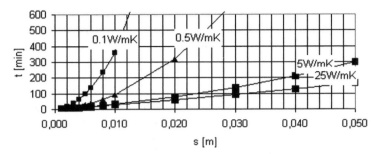

Fig. 4.7. α = 250 W/m²K, high value for free convection of a liquid or lower value for forced convection.

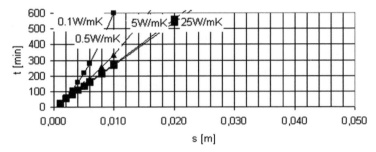

Fig. 4.8. α = 25 W/m²K, typical for forced convection of a gas or a very low value for free convection of a liquid.

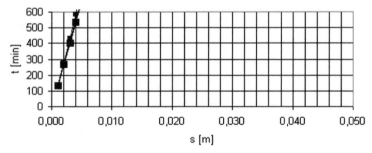

Fig. 4.9. α = 5 W/m²K, typical for free convection of a gas.

Some interesting observations are:

- Between the times calculated for the ideal case with α → ∞ (fig.4.5), that is the solution of the Stefan problem, and the times calculated for α = 2500 W/m²K (fig.4.6), like for forced convection of a liquid, there is practically no difference. For lower α the difference to the ideal case increases, especially for the PCM with high thermal conductivity.

- For the case of a liquid with forced convection with α = 250 W/m²K (fig.4.7), like in ordinary space heating systems, the time for 0.05 m is in the order of 180 min for λ = 25 W/mK (PCM-graphite matrix) .
- For the case of α = 25 W/m²K (fig.4.8), like in a liquid with free convection or a gas with forced convection, the time to reach 0.005 m typical for macroencapsulated PCM is in the order of 2 h for λ ≥ 0.5 W/mK and about 3 h for λ = 0.1 W/mK (organic PCM).
- For the case of α = 5 W/m²K (fig.4.9), like in free convection in air, the thermal conductivity of the PCM makes no significant difference anymore. To cool down 0.001 m of PCM takes about 2 h.

4.1.3 Cylindrical and spherical geometry

Baehr and Stephan 1994 calculated solutions for cylindrical geometry and spherical geometry. In these geometries, two cases exist: PCM inside and PCM outside of the spherical or cylindrical boundary. Fig.4.10 shows the different cases with the PCM marked in dark grey and the heat transfer fluid marked in bright grey. The radius R marks the location of the inner boundary; to the outside, the extension is again to infinity.

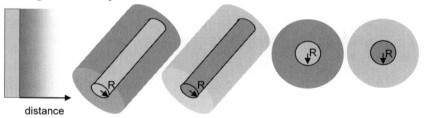

heated / cooled plane	heated / cooled cylinder	heated / cooled cylinder	heated / cooled sphere	heated / cooled sphere
PCM on one side	PCM outside	PCM inside	PCM outside	PCM inside
like in storage with flat plate heat exchanger	like in storage with pipe heat exchanger	like PCM in cylindrical encapsulation	like heating with a point source	like PCM in spherical capsules

Fig. 4.10. Plane, cylindrical, and spherical geometries that were treated by Baehr and Stephan 1994.

The solutions for these geometries can be applied to problems with pipe heat exchangers (heating / cooling cylinder, PCM outside) or cylindrical and spherical encapsulations (cylinder or sphere with PCM inside, heated / cooled from outside). Further on, the solutions give some general insight into the influence of the geometry on heat transfer. They are therefore discussed now.

Due to the cylindrical or spherical geometry a new parameter, the radius R, comes into the calculation. The time that the phase front needs to proceed a distance s away from the surface of the boundary at R, is written in the generalized form

$$t = \frac{\Delta_{pc} h \cdot s^2}{2 \cdot \lambda \cdot (T_{pc} - T_0)} f(s^+, \beta) \qquad s^+ = \frac{s}{R} \qquad (4.15)$$

It includes the special cases for the 1-dimensional problem with and without boundary effect treated by eq.4.8 and eq.4.13. The dimensionless number f, which is a function of s^+ and β, describes the deviation of the solution from the solution of the Stephan problem. Hereby, s^+ is the dimensionless ratio of the distance s of the phase front to the radius R of the cylinder or sphere. The dimensionless number β describes the influence of the boundary as it includes the parameters λ_{wall}, d_{wall}, and α with respect to R. The values for the dimensionless numbers $f(s^+,\beta)$ and β are given in tab.4.2.

Table 4.2. Solutions for cylindrical and spherical geometry, given by Baehr and Stephan 1994. For plane geometry, the values are derived here.

	$f(s^+, \beta) =$	$\beta =$
Plane, PCM one side $s^+ \geq 0$	$1 + 2 \cdot \dfrac{\lambda}{k \cdot s}$ equals $1 + 2 \cdot \dfrac{\lambda}{k \cdot R \cdot s^+} = 1 + 2 \cdot \dfrac{\beta}{s^+}$ with $\beta = \dfrac{\lambda}{R \cdot k}$	and $\beta = \dfrac{\lambda}{R \cdot k}$ equals $\beta = \dfrac{\lambda}{R} \cdot \dfrac{1}{k}$ $= \dfrac{\lambda}{R} \cdot \left(\dfrac{1}{\lambda_{wall}/d_{wall}} + \dfrac{1}{\alpha_{fluid}} \right)$ $= \dfrac{\lambda}{\lambda_{wall}} \cdot \dfrac{1}{R/d_{wall}} + \dfrac{\lambda}{\alpha_{fluid} \cdot R}$
Cylinder, PCM outside $s^+ \geq 0$	$\left(1 + \dfrac{1}{s^+}\right)^2 \ln(1+s^+) - \left(1 + \dfrac{2}{s^+}\right)\left(\dfrac{1}{2} - \beta\right)$	$\dfrac{\lambda}{\lambda_{wall}} \ln \dfrac{R}{R - d_{wall}} + \dfrac{\lambda}{\alpha(R - d_{wall})}$
Cylinder, PCM inside $1 \geq s^+ \geq 0$	$\left(1 - \dfrac{1}{s^+}\right)^2 \ln(1-s^+) - \left(1 - \dfrac{2}{s^+}\right)\left(\dfrac{1}{2} + \beta\right)$	$\dfrac{\lambda}{\lambda_{wall}} \ln \dfrac{R + d_{wall}}{R} + \dfrac{\lambda}{\alpha(R + d_{wall})}$

Sphere, PCM outside $s^+ \geq 0$	$1+\dfrac{2}{3}s^+ +\dfrac{2\beta}{s^+}\left(1+s^+ +\dfrac{s^{+2}}{3}\right)$	$\dfrac{\lambda}{\lambda_{wall}} \cdot \dfrac{d_{wall}}{R-d_{wall}}$ $+\dfrac{\lambda}{\alpha(R-d_{wall})} \cdot \dfrac{R}{R-d_{wall}}$
Sphere, PCM inside $1 \geq s^+ \geq 0$	$1-\dfrac{2}{3}s^+ +\dfrac{2\beta}{s^+}\left(1-s^+ +\dfrac{s^{+2}}{3}\right)$	$\dfrac{\lambda}{\lambda_{wall}} \cdot \dfrac{d_{wall}}{R+d_{wall}}$ $+\dfrac{\lambda}{\alpha(R+d_{wall})} \cdot \dfrac{R}{R+d_{wall}}$

For cylindrical and spherical geometry β and f were calculated by Baehr and Stephan 1994. For plane geometry, the values have been derived here by a transformation of the solution from Baehr and Stephan 1994 (eq.4.13) into the notation of tab.4.2.

It is interesting to look at the extreme case where the thermal resistance of the wall and fluid are negligible. For small wall thickness $d_{wall} \ll R$ and $\lambda_{wall} \gg \lambda$ of the PCM, the first term of β becomes zero and $\beta = \lambda / (\alpha \cdot R)$ in all cases. The temperature drop between the boundary and T_0 vanishes completely if additionally $\alpha \cdot R \gg \lambda$; then β becomes zero and the result is shown on the left in tab.4.3. For the case of the plane, with PCM on one side, the limit of negligible boundary effects is equal to the solution of the Stephan problem ($f = 1$). For the cylindrical and spherical geometries, the solutions still look somewhat complicated.

For this special case of negligible boundary effect (β = 0), it is interesting to look at two additional limiting cases.

The first one is the case of small distance s. It is completely logic under these assumptions that $f = 1$, because for negligible wall effects and small s the problem is practically a layer of PCM that is so thin that the curvature due to the special geometry becomes negligible. That means for β = 0 and for small distances ($s^+ \to 0$) all solutions approach $f(s^+, \beta) = 1$ and the solution is again the solution of the Stephan problem

$$t = \frac{\Delta_{pc} h \cdot s^2}{2 \cdot \lambda \cdot (T_{pc} - T_0)} \cdot f(s^+, \beta) = \frac{\Delta_{pc} h \cdot s^2}{2 \cdot \lambda \cdot (T_{pc} - T_0)} \tag{4.16}$$

The second case to look at is for large s ($s^+ \to \infty$). Here there are only three possible geometries to treat, because the case of large s is only possible when the PCM is on the outside. The solutions for the three cases are listed in tab.4.3, at the right column.

Table 4.3. Solutions for eliminated boundary effects ($\beta = 0$), and for large distances to the boundary ($s^+ \to \infty$)

	$f(s^+, \beta = 0) =$	$f(s^+ \to \infty, \beta = 0) =$
Plane, PCM one side	1	1
Cylinder, PCM outside $s^+ \geq 0$	$\left(1+\dfrac{1}{s^+}\right)^2 \ln(1+s^+) - \left(1+\dfrac{2}{s^+}\right)\left(\dfrac{1}{2}\right)$	$\ln(s^+) - \dfrac{1}{2} = \ln\left(\dfrac{s}{R}\right) - \dfrac{1}{2}$ $= \ln\left(\dfrac{s}{R}\right)$
Cylinder, PCM inside $1 \geq s^+ \geq 0$	$\left(1-\dfrac{1}{s^+}\right)^2 \ln(1-s^+) - \left(1-\dfrac{2}{s^+}\right)\left(\dfrac{1}{2}\right)$	
Sphere, PCM outside $s^+ \geq 0$	$1 + \dfrac{2}{3}s^+$	$\dfrac{2}{3}\dfrac{s}{R}$
Sphere, PCM inside $1 \geq s^+ \geq 0$	$1 - \dfrac{2}{3}s^+$	

For large distances $s^+ \to \infty$, all solutions are different with respect to s. Fig.4.11 shows the time-distance relation $t \sim s^2 \cdot f(s^+, \beta)$ for $R = 1$; s grows the fastest for the heating or cooling from a plane, then from a cylinder, and the slowest away from the sphere.

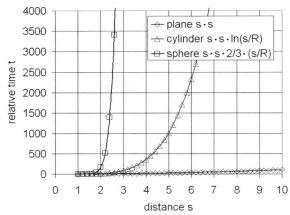

Fig. 4.11. Time-distance relation for the limit of distance $s \gg R$. As s is proportional to the power, the power of the spherical geometry is smoother than the cylindrical geometry, and that one is much smoother than the plane geometry.

This has a significant consequence on the heat transfer: the faster the change in s, the faster the change of the heat flux dQ/dt with time. Which heat flux is larger is however not straight forward to answer. The reason is that eq.4.4 is not valid in cylindrical or spherical geometry.

Fig. 4.12. Geometry and accessible volume.

Fig.4.12 shows cubes with the same volume of storage material. The left one is heated or cooled by a plane, which divides the cube into parts of identical size. The cube at the middle is heated or cooled from a cylinder at its centre, the cube at the right by a sphere at its centre. In all cases, the phase front has to move a distance equal the half of the cube dimension to reach the cubes surface. The storage volume in the left cube has the largest size and the phase front reaches the boundaries the fastest according to fig.4.11. At the centre, the cylinder does not fill the cube completely. Additionally, it takes longer for the phase front to reach the cubes surface than for the plane geometry. In the spherical geometry, it takes the longest for the phase front to reach the cubes surface and the volume of storage material reached is even smaller than for the cylindrical geometry. Therefore, the power of heating or cooling decreases from plane geometry to cylindrical geometry and to spherical geometry.

It is also interesting to look at the extreme case where the thermal resistance of the wall and fluid are dominating, that means for large β. The solutions are listed in tab.4.4.

Table 4.4. Comparison of the general case (left) to the case for dominating boundary effects (right).

	$f(s^+,\beta) =$	$f(s^+,\beta \to \infty) =$
Plane, PCM one side $s^+ \geq 0$	$1 + 2 \cdot \dfrac{\beta}{s^+}$	$2 \cdot \dfrac{\beta}{s^+}$
Cylinder, PCM outside $s^+ \geq 0$	$\left(1+\dfrac{1}{s^+}\right)^2 \ln(1+s^+) - \left(1+\dfrac{2}{s^+}\right)\left(\dfrac{1}{2}-\beta\right)$	$2 \cdot \dfrac{\beta}{s^+} + \beta$
Cylinder, PCM inside $1 \geq s^+ \geq 0$	$\left(1-\dfrac{1}{s^+}\right)^2 \ln(1-s^+) - \left(1-\dfrac{2}{s^+}\right)\left(\dfrac{1}{2}+\beta\right)$	$2 \cdot \dfrac{\beta}{s^+} - \beta$

(Continued)

Table 4.4. (Continued)

	$f(s^+,\beta) =$	$f(s^+,\beta \to \infty) =$
Sphere, PCM outside $s^+ \geq 0$	$1 + \dfrac{2}{3}s^+ + \dfrac{2\beta}{s^+}\left(1 + s^+ + \dfrac{s^{+2}}{3}\right)$	$2 \cdot \dfrac{\beta}{s^+}\cdot\left(1 + s^+ + \dfrac{s^{+2}}{3}\right)$
Sphere, PCM inside $1 \geq s^+ \geq 0$	$1 - \dfrac{2}{3}s^+ + \dfrac{2\beta}{s^+}\left(1 - s^+ + \dfrac{s^{+2}}{3}\right)$	$2 \cdot \dfrac{\beta}{s^+}\cdot\left(1 - s^+ + \dfrac{s^{+2}}{3}\right)$

For small distances s compared to R additionally s^+ becomes small and in all cases $f(s^+, \beta \to \infty)$ approaches the same value $2 \cdot \beta / s^+$. This is logical, because for s very small compared to R the effect of the geometry has to vanish.

4.1.4 Layer with finite thickness

When sensible heat is neglected, it is also possible to treat layers with finite thickness, even when heat is extracted or supplied from two sides.

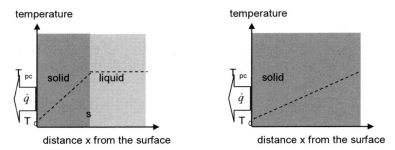

Fig. 4.13. Cooling of a layer of finite thickness from one side until the phase front reaches the rear end of the layer.

Fig.4.13 shows the cooling of a layer with finite thickness. Because the calculations are based from the beginning on the assumption that the temperature is at the melting temperature throughout, the effect of the finite thickness does not have any effect until the phase front at s has reached the rear end of the PCM layer. The formulas can therefore be used as before, it is only necessary to check that s does not exceed the thickness of the layer. When cooling a layer from both sides, the situation is very similar (fig.4.14).

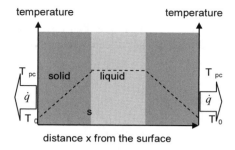

Fig. 4.14. Cooling of a layer of finite thickness from both sides.

Now the phase front proceeds into the layer from both sides. The two different phase fronts will meet somewhere and the cooling is finished. It is only necessary to check that the sum of both distances of the phase fronts from the surfaces does not exceed the thickness of the layer. The same approach can also be used for cylindrical and spherical problems.

4.1.5 Summary and conclusion for analytical models

The models for the heat transfer described in this chapter have rather simple analytical solutions. These solutions give general insight into the basic relations between the different parameters T_{pc}, T_0, λ, $\Delta_{pc}h$, and t on the variation of dQ/dt and s with time. Probably the most important one is the relation between the location of the phase front at s and the time t. The different geometries resulting from an encapsulation or from a heat exchanger can lead to completely different variation of s with time, as derived in section 4.1.3

Analytical solutions can give an important hint if for example encapsulations as flat plates, pipes, or spheres should be used in an application with respect to the desired time development of the heat flux. The location of the phase front s with time relates the heat flux to the geometric dimensions. The largest geometric dimension for the phase front to travel determines the average and final heat flux before the phase change is complete.

Despite this important, general information given by analytical solutions, it is important to keep in mind that strong limitations exist with respect to geometry and, what is more important, to thermal effects. These restrictions often lead to poor results when using analytical models to get quantitative information. Analytical models do not include sensible heat, only treat heat conduction and not convection, assume that the PCM is at the phase change temperature at the beginning, and that boundary temperatures are constant. Another restrictive limitation is the assumption of a sharp melting temperature. Most PCM have a melting range several K wide. If the temperature at the boundary comes close to the melting range, this assumption leads to significant errors. If the temperature of the boundary is very different from the melting range, this restriction is negligible. However, then the sensible heat becomes significant and cannot be neglected any more.

4.2 Numerical models

For a detailed and more realistic analysis of real problems, a different approach is necessary: the use of numerical models. In numerical models, it is only necessary to define the problem by differential equations. The problem of solving the differential equation is then left to the computer. Numerical models are therefore more flexible with respect to thermal effects, and to the geometry that can be treated.

In steady state heat transfer problems, the differential equations have to be solved in the space of interest. All heat storage problems are however dynamic problems; therefore it is additionally necessary to find the solution in time. To show how numerical models work the discussion is again started with a heat transfer problem in 1-dimension.

4.2.1 1-dimensional PCM layer

The problem of time dependent heat transfer consists of two steps:

1. The heat transfer between the different area or volume elements
2. The change in temperature within each area or volume element due to that heat transfer

To get the solution of the heat transfer problem, that is the time variation of the temperatures and heat flux, these two steps have to be calculated repeatedly. The two steps are now discussed in more detail.

Heat transfer between different areas or volume elements

For simplicity, the discussion will treat only heat conduction. To treat additionally convection and radiation, the equations and their solutions are more complicated. Further on, it is necessary to know additional parameters like viscosity and density. These however are usually not available for PCM in the melting range.

The heat flux by heat conduction is given by Fourier's law. For a 1-dimensional problem, it can be written as

$$\dot{Q} = -A \cdot \lambda \cdot \frac{dT}{dx}. \tag{4.17}$$

To describe the heat transfer in space, the space is divided into different areas or volume elements, as shown in fig.4.15. For simplicity, they all have the same size: the division into Δx is equal to a division into volume elements with $\Delta x \cdot A = \Delta V$. Hereby A is the area that represents the extension into the remaining two space coordinates.

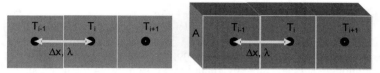

Fig. 4.15. Division of the space coordinate into finite areas or volume elements denoted by the variable i.

To do numerical calculations systematically, each area becomes a number denoted by the variable i, here as subscript. The physical properties of the material in the calculation are represented by a node for each volume element. To visualize that representation, the nodes lie in the center of each volume element. T_i for example denotes the temperature T at node i.

In a heat transfer problem with heat conduction only, the thermal properties of a node are its temperature, heat storage capacity, and thermal conductivity. According to fig.4.15 and eq.4.17, the heat ΔQ transferred between the two adjacent nodes i and i-1 in a time interval Δt is

$$\frac{\Delta Q}{\Delta t} = -A \cdot \lambda \cdot \frac{T_i - T_{i-1}}{x_i - x_{i-1}}. \tag{4.18}$$

Here, A is the heat exchanging area between the two nodes, λ the thermal conductivity of the material, $\Delta T = T_i - T_{i-1}$ the temperature difference, and $\Delta x = x_i - x_{i-1}$ the distance between the two nodes. In eq.4.18 the differential equation from eq.4.17 becomes a finite difference equation. In a *finite difference method*, the differentials in a differential equation are replaced by finite a difference, which means the differential of the heat dQ is replaced by ΔQ, and for t, T, x the same way.

Change in temperature due to heat transfer

The amount of heat transferred between two nodes in a time interval Δt is

$$\Delta Q = -\Delta t \cdot A \cdot \lambda \cdot \frac{\Delta T}{\Delta x}. \tag{4.19}$$

The heat transferred will lead to a temperature change of the nodes. At this point, it is necessary to introduce a second subscript for the time, using the variable j and the time interval Δt. In numerics, it is often said that the parameters i with distance Δx and j with Δt form the grid for the numerical calculation.

In the 1-dimensional problem, a volume element usually has two neighbor elements. The heat transferred into volume element i from the two neighbor elements i + 1 and i - 1 during time step j is then

$$\Delta Q_{i,j} = \Delta t \cdot A \cdot \lambda \cdot \frac{T_{i+1,j} - T_{i,j}}{\Delta x} - \Delta t \cdot A \cdot \lambda \cdot \frac{T_{i,j} - T_{i-1,j}}{\Delta x}$$

$$= \frac{\Delta t \cdot A \cdot \lambda}{\Delta x} \cdot \left(T_{i+1,j} - 2 \cdot T_{i,j} + T_{i-1,j}\right) \qquad (4.20)$$

Hereby it is taken into account that from node i+1 to i the direction is the opposite than from node i-1; this is done by changing the sign. From the heat gain or loss in one time step, it is possible to calculate the new temperatures in the next time step. To do this it is necessary to make a mathematical connection between the heat storage capacity described by c_p or h and the temperature T. Fig.4.16 shows different possible functions describing this connection for the case of a melting range and for the case of a melting temperature.

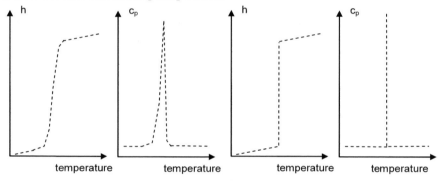

Fig. 4.16. Heat storage capacity described by c_p or h: on the left with a melting range, which is typical for many materials; on the right with a melting temperature, typical for pure materials.

Description of temperature change of a node using the heat capacity c_p

Using the heat capacity $C_p = \Delta V \cdot c_p$ of a node, the new temperature of a node at time step j+1 can be calculated from the heat lost or gained by the node and its initial temperature at time step j, by

$$\Delta Q_{i,j} = \Delta H_{i,j} = \Delta V \cdot c_p(T_{i,j}) \cdot \left(T_{i,j+1} - T_{i,j}\right). \qquad (4.21)$$

The value of c_p here has to be $c_p(T_{i,j})$ exactly speaking. From eq.4.20 and eq.4.21 follows that

$$\Delta V \cdot c_p(T_{i,j}) \cdot (T_{i,j+1} - T_{i,j}) = \frac{\Delta t \cdot A \cdot \lambda}{\Delta x} \cdot (T_{i+1,j} - 2 \cdot T_{i,j} + T_{i-1,j}). \tag{4.22}$$

Replacing ΔV by $A \cdot \Delta x$ gives

$$c_p(T_{i,j}) \cdot (T_{i,j+1} - T_{i,j}) = \frac{\Delta t \cdot \lambda}{\Delta x^2} \cdot (T_{i+1,j} - 2 \cdot T_{i,j} + T_{i-1,j}). \tag{4.23}$$

Separating the temperature of the new time step to the left side of the formula, the equation becomes

$$T_{i,j+1} = T_{i,j} + \frac{\Delta t \cdot \lambda}{c_p(T_{i,j}) \cdot \Delta x^2} \cdot (T_{i+1,j} - 2 \cdot T_{i,j} + T_{i-1,j}). \tag{4.24}$$

This equation allows to calculate the new temperature at time step j + 1 of volume element i from known values at the last time step j. The calculations have to be done by a computer successively for all volume elements and give the solution for the time dependent heat transfer.

In eq.4.24, all input data are known from the earlier time step. This way of calculating the values for the next time step only from information from the preceding time steps is called *explicit method*. To improve the numerical stability of a calculation one can also use an *implicit method*. In an implicit method, values of the new time step for all nodes are calculated at the same time. The programming here is however more complicated. A more detailed treatment of numerical modeling is in Farlow 1982 and Özisik 1968.

Compared to analytical models, numerical models have significant advantages. The restriction on a simple geometry that was necessary to derive analytical models can now simply be resolved by using a more complex structure of nodes. The restrictions 1, 3, and 4 for analytical models in section 4.1.1, namely

- heat is stored only as latent heat,
- at t = 0 the PCM is completely liquid and exactly at the phase change temperature throughout, and
- the temperature at x = 0 is changed at t = 0 to T_0 and kept at that level

can now easily be resolved using a real $c_p(T)$-function like in fig.4.16 and by defining a temperature-time curve at the boundary where the heat exchanger is placed. At other boundaries, similar functions can be applied to include heat losses or adiabatic conditions. The restriction number 2, that is heat transfer is only by conduction, is more difficult to resolve. This case is not discussed here because its treatment goes beyond an introduction. Interested readers find literature references at the end of this chapter.

The approach to describe the temperature change of nodes using the heat capacity c_p in eq.4.21 is straight forward, but leads to a problem. In eq.4.24, the new temperature for time step j + 1 at a node i is calculated using the heat capacity at the temperature at time step j. If the time step is chosen too large, the enthalpy difference between the old and the new temperature will be wrong.

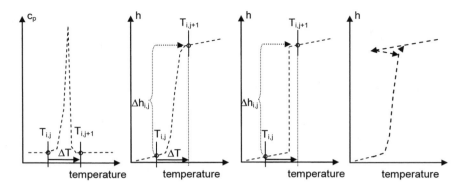

Fig. 4.17. Left: example where using the heat capacity leads to a significant error. Centre to right: using the enthalpy a melting rage, a melting temperature, and even subcooling can be modeled.

Fig.4.17 shows an extreme case, where the new temperature at time step $j + 1$ is on the other side of the melting peak. In that case, the melting enthalpy would not be included in the calculation at all, because the enthalpy difference is the temperature change multiplied by the heat capacity at the initial time (eq.4.21). The simulation of a rapid changing heat capacity requires therefore very small time steps and the influence of the time step should be checked on the result of a numerical model.

Description of temperature change of a node using the enthalpy

This problem can be avoided when the mathematical connection between the heat storage capacity and the temperature is done using the enthalpy h; the corresponding method is called the *enthalpy method*. The description of heat transfer between nodes is still done using eq.4.20, but instead of relating the change of heat ΔQ to a temperature change by the heat capacity as in eq.4.21 it is now done by the enthalpy

$$\Delta Q_{i,j} = \Delta H_{i,j} = \Delta h_{i,j} \cdot \Delta V = \frac{\Delta t \cdot A \cdot \lambda}{\Delta x} \cdot \left(T_{i+1,j} - 2 \cdot T_{i,j} + T_{i-1,j}\right). \quad (4.25)$$

Substitution of $\Delta V = A \cdot \Delta x$ leads to

$$\Delta h_{i,j} = \frac{\Delta t \cdot \lambda}{\Delta x^2} \cdot \left(T_{i+1,j} - 2 \cdot T_{i,j} + T_{i-1,j}\right). \quad (4.26)$$

Finally

$$h_{i,j+1} = h_{i,j} + \frac{\Delta t \cdot \lambda}{\Delta x^2} \cdot \left(T_{i+1,j} - 2 \cdot T_{i,j} + T_{i-1,j} \right) \tag{4.27}$$

The new temperature at time step $j + 1$ can now be determined from the h(T) function as shown in fig.4.17. The case of a melting range or of a single melting temperature can be treated without any problem.

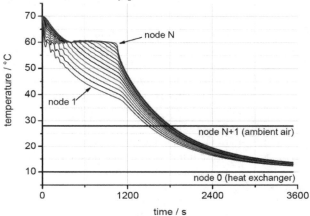

Fig. 4.18. Result of a numerical simulation for 1-dimensional cooling (picture: ZAE Bayern).

Fig.4.18 shows the result of a numerical simulation for 1-dimensional cooling with N=15 nodes. The first node, which is closest to the cooling surface of the heat exchanger, cools down very fast. Nodes that are more distant cool down slower and node 5 and higher show a plateau in the temperature history at the phase change temperature. Because sensible heat is included in the simulation, the temperature drop after the phase change is completed is slowed down and the temperatures slowly approach the temperature of the heat exchanger. It is also possible from these data to get a qualitative idea of the cooling power. The heat transferred to the heat exchanger is equal to the temperature gradient at the heat exchanger, which is the temperature gradient between nodes 0 and 1, multiplied by the area and the thermal conductivity. Because area and thermal conductivity are constant, the heat transferred to the heat exchanger is proportional to the temperature difference between nodes 1 and 0. Usually people expect to see a plateau in the heat transferred to the heat exchanger also, but the numerical results show that this is not the case. The reason is that even though the heat is mainly released at a constant temperature, the phase change temperature, the thermal resistance to the heat exchanger is constantly changing with the distance of the phase front to the heat exchanger.

4.2.2 Inclusion of subcooling using the enthalpy method

Subcooling is the effect that a temperature significantly below the melting temperature has to be reached, until a material begins to solidify and release heat (fig.2.1). Fig.2.19 shows the subcooling of NaOAc·3H$_2$O with a nucleator added; crystallization started at a temperature of 2 K below the melting temperature. Without nucleator, subcooling is usually much stronger. If the temperature where crystallization starts is not reached, the PCM will not solidify at all and thus only store sensible heat. When the temperature changes in an application are small, even little subcooling can have a mayor effect on the performance of the system.

In state of the art simulations of heat transfer with phase change, the effect of subcooling is usually neglected. This is a practicable approach if subcooling is small compared to the temperature changes in an application, but for significant subcooling, the results can be completely wrong. In such cases, subcooling has to be included into the numerical model. None of the models described above includes subcooling, because these models are based on the assumption that for each value of the enthalpy h there is one, and only one, defined value of the temperature T. During subcooling, as shown in fig.2.17 on the right, there are however two values. It is straight forward therefore to use two h(T) relations to include subcooling into the model. Upon cooling, the "cooling" h(T) relation is used until a temperature is reached where solidification starts and then the "heating" h(T) relation is used again. For the change between both relations, some authors have used the approach that it is done for all nodes, that means throughout the whole volume, at the same time. This means that nucleation starts everywhere at the same time. In reality, the solid phase formed will nucleate only the adjacent PCM at a speed determined by the growth rate of the crystal and the rate of heat dissipation. The approach to do the change between both h(T) relations everywhere at the same time is therefore only correct when this speed is high enough to reach the whole volume within a single time step.

A numerical procedure to solve this problem has been developed at the ZAE Bayern within the project LWSNet (funded by the German Federal Ministry of Education and Research, BMBF). Günther et al. 2007 have presented the procedure and its verification by a comparison to experiments. In order to take account the effect of subcooling, the approach of using different h(T) relations just described is used. However, it is modified to take into account the crystallization speed correctly. The ambiguity of h(T) is solved by introducing a second parameter, namely the phase p, such that h = h(T, p). This allows the treatment of different cases at the same temperature, for example p = 0 in the liquid state and p = 1 in the solid state. Upon cooling, the phase remains liquid until the crystallization is triggered. This triggering in one volume element starts when the temperature in the volume element falls below the nucleation temperature or by direct contact to an adjacent volume element that is already crystallized. The correct treatment of the second effect is critical when the bulk of the material is cooled down at a significant speed, because different nodes will have significantly different temperatures at the same time. For example, in the 1-dimensional problem shown in fig.4.2 the surface is cooled and nucleation will start first where the coldest nodes are. This is at the surface at x = 0. If the speed of crystallization is fast enough, nodes that are more distant will not experience any subcooling. If the simulation uses an explicit method, the information that nucleation is triggered is spread by a speed which is determined automatically by the time step. At each time step all neighbors of already nucleated nodes would also crystallize. In the new algorithm, the speed of solidification is controlled artificially by introducing the degree of solidification of a node. If the phase change is triggered in a node, its phase parameter p is at first set to the value 2, representing a mixed liquid-solid node. In each time step, the new degree of solidification is determined according to the node size and the speed of solidification. Only upon completion of the phase change, the phase parameter of the node is set to 1, representing a solid node. Consequently, during the solidification of one node, it cannot trigger crystallization of an adjacent node, and the speed of phase change is controlled.

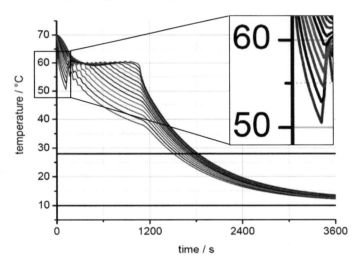

Fig. 4.19. Result of a numerical simulation for 1-dimensional cooling including the effect of subcooling (picture: ZAE Bayern).

Fig.4.19 shows the result of a numerical simulation for 1-dimensional cooling including the effect of subcooling. Compared to fig.4.18 a significant change in the temperature evolution of the first nodes is visible. The first node cools down the fastest, reaches the nucleation temperature first, and starts to crystallize. Nodes that are more distant follow later. The effect of nucleation and of a fixed crystal growth rate has significant consequences that are also visible. The nucleation by already solidified nodes leads to smaller subcooling in adjacent nodes, getting smaller with larger distance to the heat exchanger. The enlarged section shows that the effect of a fixed crystal growth rate leads to a time delay between the nucleation of different nodes, getting later with larger distance to the heat exchanger.

4.2.3 Relation between h(T) functions and phase diagrams

Now that the discussion on how to include arbitrary h(T) functions in numerical models is complete, there is an important question remaining: what is the origin of the shape of the function? More specific, what is the relation between the h(T) functions and phase diagrams.

For the case of a pure substance, like water, the answer is simple: the phase diagram (fig.2.11) shows a melting temperature, and as a consequence the h(T) function has a step at the melting temperature. This case is shown on the left side in fig.4.20. T_0 is a reference temperature.

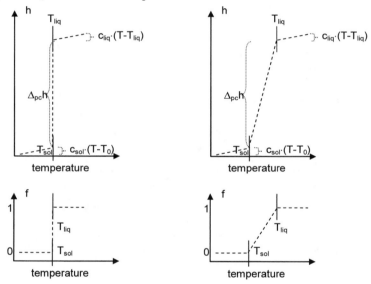

Fig. 4.20. Left: h(T) function with a step at the melting temperature. Right: h(T) function with a linear rise in a melting range. The parameter f is the fraction of the phase change enthalpy $\Delta_{pc}h$ to take into account in h.

The graph at the bottom in fig.4.20 shows how the fraction f of the phase change enthalpy $\Delta_{pc}h$ in h changes with temperature. The fraction f is equal to zero when the material is completely solid and equal to one when the material is completely liquid. Therefore, f is often associated with the fraction of the liquid phase.

In the same way as a pure substance like water is treated, salt hydrates with congruent melting (fig.2.13) and eutectic mixtures (fig.2.14) can be treated. The reason is that they also show a melting temperature and a phase change between a homogeneous liquid phase and a homogeneous solid phase. In the case of a noneutectic water-salt solution, the situation is different. Due to the noneutectic composition, there is no sharp melting temperature. An example of a phase diagram is shown on the left in fig.4.21. At about -13 °C water freezes to ice; the remaining liquid solution carries the salt and thus increases in salt concentration. At a temperature of about -27 °C, the eutectic point is reached and the remaining salt solution with a salt concentration of about 15 wt.% solidifies. Because the phase change takes place in a temperature range, the corresponding h(T) function must take this effect into account, as shown on the right in fig.4.20. The graph at the bottom in fig.4.20 shows again, how the fraction f of the phase change enthalpy $\Delta_{pc}h$ changes with temperature. However now it is not completely correct to use the value of f also as the fraction of the liquid phase. The reason is that the composition of the liquid phase changes with temperature and thereby changes the thermal effect when a new portion of ice is formed.

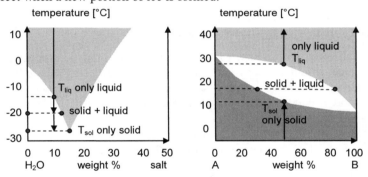

Fig. 4.21. Left: phase diagram of a noneutectic water-salt solution. Right: phase diagram typical for alkane mixtures.

Another example shows the phase diagram on the right in fig.4.21. It describes a 2-component mixture with a composition of 50 wt.% each. An example could be the mixture of two alkanes. The mixture is solid with one component dissolved in the other one up to T_{sol}. At T_{sol} the mixture starts to melt. During melting, a solid and a liquid fraction form. In the process of melting, the compositions of the solid and the liquid continuously change with rising temperature. When the temperature is raised above T_{liq}, the mixture is completely liquid and homogeneous with the composition of 50 - 50 wt.% again. The melting range of the mixture is the temperature interval between T_{sol} and T_{liq}. It is again described by a h(T) function as shown on the right in fig.4.20. As was the case with noneutectic water-salt solutions, f cannot directly be associated with the fraction of the liquid phase. The reason is that the composition of the liquid phase, and now also of the solid phase, changes with temperature and thereby also changes the thermal effect when a new portion of solid or liquid phase is formed. Upon solidification, there is an additional effect due to the kinetics of diffusion: the diffusion of molecules from the liquid into the solid to reach the correct composition takes much more time than diffusion in the liquid phase. This can significantly slow down the formation of the solid phase with the correct composition, and consequently slow down the release of the phase change enthalpy. For semi congruent or incongruent melting PCM the situation can be even more complex.

For a numerical calculation with the enthalpy method, it is necessary to know the enthalpy h as a function of the temperature. For pure substances, the enthalpy can be calculated from literature values of the heat capacity c of the solid and liquid phase, and the phase change enthalpy $\Delta_{pc}h$. As a reference temperature T_0 it is convenient to use a temperature where the PCM is still solid. Because in a numerical simulation only enthalpy changes are used, the choice of the reference temperature however has no influence on the result. Taking into account that for a pure substance $T_{sol}=T_{liq}=T_{pc}$ the enthalpy is (fig.4.20)

- for $T<T_{pc}$: $h(T)=c_{sol}\cdot(T-T_0)$
- for $T=T_{pc}$: $h(T)=c_{sol}\cdot(T_{pc}-T_0)+f\cdot\Delta_{pc}h$
- for $T>T_{pc}$: $h(T)=c_{sol}\cdot(T_{pc}-T_0)+\Delta_{pc}h+c_{liq}\cdot(T-T_{pc})$

The same approach can be used to calculate the h(T) function for the case of a melting range, assuming that f is linear in the melting range. This is however a simplification and usually not very accurate. In most cases it is more convenient and also more accurate to use measured heat capacity data to determine the h(T) function by integration (eq.3.7).

The existence of a melting range also has an influence on the melting front. It extends over the melting range and is often called *mushy zone* (fig.4.22).

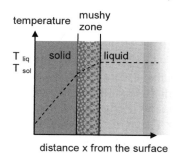

Fig. 4.22. Melting front, often called mushy zone, in a cooling process.

Besides the heat storage capacity, other properties like viscosity and thermal conductivity (fig.3.27) vary within the melting range. It must be pointed out that in contrast to the enthalpy, the overall thermal conductivity and viscosity are usually not equal to the average of the thermal conductivity of the solid and liquid phase weighed by their respective fractions.

4.3 Modellization using commercial software

The simulation of heat transfer in a PCM in a simple geometry using the approach described in section 4.1 and 4.2 can be done with many mathematical and engineering software tools like MathCad, Mathematica, and EES. If someone does not like to spend time on doing the mathematics and the numerics, results can be achieved with commercial CFD (computational fluid dynamics) software solutions. Some of them can handle phase changes like FLUENT and ANSYS CFX11, however not subcooling. In any case, it is advisable to test the software, commercial or self programmed, at least on standard problems like the Stefan problem.

The main advantage using commercial software is that only very basic knowledge on numerical methods for heat transfer is required. The disadvantages are however also significant: it is difficult to understand what the software does, one usually does not gain deeper understanding, and often the accuracy is highly overestimated because of the colorful and fancy graphs. This is especially the case when convection is treated. Heat transfer by convection with PCM is different from ordinary convection. PCM can transport significant amounts of latent heat in the melting temperature range with comparatively little fluid movement and the density changes driving convection are much stronger. Currently, sufficiently accurate data of the density, viscosity, and enthalpy of PCM as a function of temperature are very rare. For example for the density, usually only one value for the liquid and one for the solid state are available. The consequence is that even with commercial software the lack of the necessary experimental data causes significant uncertainties in the results. Another big disadvantage of using commercial software becomes clear when there is a large difference in the scale of a problem. An example could be the design of a storage for house heating. To simulate the heat transfer within the storage between PCM and heat exchanger it is possible to use a CFC software. Further on, it is necessary to find out the heat demand in the building. However, while the storage heats the building the heat demand of the building is changed, therefore there is a feedback loop. For a consistent dynamic treatment, it would be necessary to integrate the solution of the CFC software for the storage into a building simulation software. This is however usually impossible because of the different numerical procedures used for the simulation of the large scale (house) and small scale (heat transfer PCM-heat exchanger). If the boundary conditions determined from the house are simple, they can be used as boundary conditions to simulate the storage with a CFD software. If the boundary conditions are complicated, and the storage shows a feed back onto its boundary conditions, then both systems have to be included in one simulation. Sometimes people try to solve the problem by describing the storage with a set of simple functions, for example giving the heating or cooling power as a function of the inlet temperature and the volume flow. This approach is however only possible for the very special case of a fully charged / discharged storage which is discharged / charged with a constant inlet temperature and volume flow. If any of the boundary conditions is changed, the performance will not be described by a function anymore. At that point, it is necessary to go back to do the numerics and write a special numerical procedure for the storage and integrate it into the building simulation.

4.4 Comparison of simulated and experimental results

The analytic and numerical models described before can be programmed quite easily, and then be used to study the effect of different parameters. After all the assumptions made, it is however necessary to test their accuracy. Do they agree with experimental data also quantitatively or only qualitatively? Besides restrictions due to the mathematical or numerical model, it is necessary to take into account that simulating phase change involves several additional variables compared to ordinary heat transfer, maybe with significant convection or subcooling, and that the material data basis is often poor with respect to availability and accuracy. To answer these questions, experiments in simple geometry have to be made and the results have to be compared with calculations based on thermodynamic property data of the PCM. For the comparison of simulation and experiment in heat transfer, it is straightforward to start again with a 1-dimensional problem.

4.4.1 1-dimensional PCM layer without subcooling

For the experimental investigation of the heating and cooling of a PCM layer the set-up shown in fig.4.23 was developed. It consists of a heat exchanger attached to a container that holds the PCM.

Fig. 4.23. Experimental set-up to investigate the heating (right) and cooling (left) of a semi-infinite PCM layer (picture: ZAE Bayern).

The heat exchanger is made from a copper block and is connected to a thermostat. Further on, the heat exchanger was built in a way that it does not add another thermal resistance to the calculation and that it allows a heating or cooling to a new temperature within less than a minute. The container for the PCM is made of acrylic glass to allow visual observation of the PCM during the experiments. Within the container, thermocouples are located at the surface of the heat exchanger and in different distances to record the temperature evolution with time.

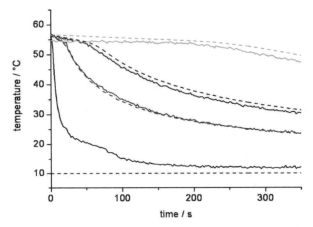

Fig. 4.24. Results of experiment (—) and simulation (- -) for the temperature-time history in a 1-dimensional cooling experiment (picture: ZAE Bayern).

Fig.4.24 shows the temperature-time history in a cooling experiment using a PCM with a melting temperature at about 54 °C. After melting the PCM at a temperature that was only slightly higher than the melting temperature, the temperature at the heat exchanger was changed rapidly to about 10 °C. A comparison of the experimental and simulated temperature curves shows that using a suitable numerical model and the correct thermophysical properties a good agreement can be achieved. The deviation at the heat exchanger is due to the fact that the thermocouple is mounted at the boundary between heat exchanger and sample. Therefore, the thermocouple can not measure the surface temperature of the heat exchanger correctly.

4.4.2 1-dimensional PCM layer with subcooling

The numerical model that includes subcooling described above has also been verified experimentally. Fig.4.25 shows the comparison of simulated and experimental data for the cooling of a NaOAc·3H$_2$O sample. This PCM was chosen because significant subcooling can be achieved.

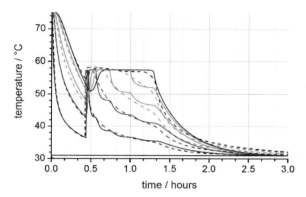

Fig. 4.25. Results of simulation (—) and experiment (- -) for subcooling of NaOAc·3H$_2$O (picture: ZAE Bayern).

The results show that simulation and experiment agree quite well. The only significant differences are fluctuations in the simulated data caused by the limited resolution of the numerical grid. More details on the test and further results not described here are available in the publication by Günther et al. 2007.

4.5 Summary and conclusion

In this chapter, the basics of heat transfer in PCM have been described. Only the transfer of heat allows the storage of heat in a PCM, or the stabilization of the temperature of something else by a PCM. It was shown how to calculate the heat flux and the time to complete phase change using analytical and numerical methods, the assumptions they are based on and their restrictions, different geometries, and the inclusion of subcooling.

For people interested in analytical, numerical and experimental work in the area of phase change, especially freezing and melting processes, the reviews carried out by Eckert et al.1997 and Hu and Argyropoulos 1996 can be recommend. A general overview is the treatment by Carslaw and Jaeger 1959. Many references on how to include convection, as well as on other numerical and analytical approaches, can be found in Zalba et al. 2003. Voller 1987 for example describes an enthalpy method for convection / diffusion phase change. Ismail 2002 and Lane 1983 discuss the influence of the geometry. Dincer and Rosen 2002 describe on an advanced level heat transfer with phase change in simple and complex geometries, for cylinders, spheres, and plane geometries; fins are also treated. Further on, an extensive list of references is given. Recent works on the problem of simulation and the comparison of results with experiments are Sasaguchi and Viskanta 1989, Abugderah and Ismail 2003, Hirose et al. 2003, and Ziskind and Letan 2007.

4.6 References

[Abugderah and Ismail 2003] Abugderah M.M., Ismail K.A.R.: Comparison study between numerical and simplified numerical analysis of PCM Thermal Storage Systems. Proc. of Futurestock 2003, 9th International Conference on Thermal Energy Storage, Warsaw, POLAND (2003)

[Baehr and Stephan 1994] Baehr H. D., Stephan K.: Wärme- und Stoffübertragung. Springer Verlag, Berlin Heidelberg (1994)

[Carslaw and Jaeger 1959] Carslaw H.S. and Jaeger J.C.: Conduction of heat in solids. Oxford and the Clarendon Press, 2nd ed. (1959)

[Çengel 1998] Çengel Y.: Heat transfer – A practical approach. Mc Graw Hill (1998)

[Dincer and Rosen 2002] Dincer I., Rosen M.A.: Thermal energy storage. Systems and applications. John Wiley & Sons, Chichester (2002)

[Eckert et al. 1997] Eckert E.R.G., Goldstein R.J., Ibele W.E., Patankar S.V., Simon T.W., Strykowski P.J., Tamma K.K., Kuehn T.H., Bar-Cohen A., Heberlein J.V.R., Hofeldt D.L., Davidson J.H., Bischof J., Kulacki F.: Heat transfer – a review of 1994 literature. Int. J. Heat Trasfer **40**, 3729–3804 (1997)

[Farlow 1982] Farlow S. J.: Partial differential equations for scientists and engineers. Dover Publications (1982)

[Günther et al. 2007] Günther E., Mehling H., Hiebler S.: Modeling of subcooling and solidification of phase change materials. Modelling Simul. Mater. Sci. Eng. **15**, 879–892 (2007) doi:10.1088/0965-0393/15/8/005

[Hirose et al. 2003] Hirose K., Yoshii T., Watanabe H.: Melting heat transfer of phase change material in horizontal tubes immersed in water. Proc. of Futurestock 2003, 9th International Conference on Thermal Energy Storage, Warsaw, Poland (2003)

[Hu and Argyropoulos 1996] Hu H., Argyropoulos S.A.: Mathematical modelling of solidification and melting: a review. Modelling Simul. Matter. Sci Eng., 371-396 (1996)

[Ismail 2002] Ismail K.A.R.: Heat transfer with phase change in simple and complex geometries. In: Dinçer I. and Rosen M.A. (eds) Thermal Energy Storage. Systems and applications, pp. 337-386. John Wiley & Sons, Chichester (2002)

[Lane 1983] Lane G.A.: Solar Heat Storage: Latent Heat Material - Volume I: Background and Scientific Principles. CRC Press, Florida (1983)

[Mehling et al. 1999] Mehling H., Hiebler S., Ziegler F.: Latent heat storage using a PCM-graphite composite material: advantages and potential applications; Presented at 4th Workshop IEA ECES Annex 10 "Phase change materials and chemical reactions for thermal energy storage", Benediktbeuern, Germany, 28.-29. Oktober 1999. www.fskab.com/annex10

[Mehling et al. 2007] Mehling H, Cabeza L.F., Yamaha M.: Phase change materials: application fundamentals. In: Paksoy H.Ö. (ed.): Thermal energy storage for sustainable energy consumption – fundamentals, case studies and design, pp. 279-314. Springer, (2007), NATO Science series II. Mathematics, Physics and Chemistry – Vol. 234, ISBN 978–1–4020–5289–7

[Özisik 1968] Özisik M.N.: Boundary value problems of heat conduction. Dover Publications (1968)

[Sasaguchi and Viskanta 1989] Sasaguchi K., Viskanta R.: Phase change heat transfer during melting and resolidification of melt around cylindrical heat source(s)/ sink(s). Journal of Energy Resources Technology **111**, 43-49 (1989)

[Stefan 1891] Stefan J.: Über die Theorie der Eisbildung, insbesondere die Eisbildung im Polarmeere. Ann. Phys. U. Chem. **42**, 269-286 (1891)

[Voller 1987] Voller V.R.: An enthalpy method for convection / diffusion phase change. International Journal for numerical methods in engineering **24**, 271-284 (1987)

[Zalba et al. 2003] Zalba B., Marin J.M., Cabeza L.F., Mehling H.: Review on thermal energy storage with phase change materials, heat transfer analysis and applications. Applied thermal Engineering **23**, 251 – 283 (2003)

[Ziskind and Letan 2007] Ziskind G., Letan R.: Phase change materials: recent advances in modelling and experimentation. Proc. of Heat SET 2007, Heat Transfer in Components and Systems for Sustainable Energy Technologies, 18-20 April 2007, Chambery, France, pp.685-700 (2007)

5 Design of latent heat storages

In chapter 4, it was shown how to calculate the heat flux and the time to complete phase change in a simple geometry. In chapter 5, this knowledge is used as a basis to design complete storages. For hot water heat storages, there are many different design options. For example, heat can be extracted via a heat exchanger or by taking out the stored hot water directly. Heat can also be supplied to the storage by these two ways. The function of the heat exchanger is to separate heat source, heat storage, and heat sink from each other in order to use different media, pressure, or purity. The variety of design options for latent heat storages is even larger. The reason is that besides water as heat transfer and storage medium, the use of a PCM as storage medium adds a second material to the storage with completely different behavior and effects.

This chapter gives an introduction and an overview on design criteria and the most common design options of LHS. For each design option an example and a simplified calculation of the heat transfer is included. The chapters on different applications will discuss more examples including boundary conditions and integration into energy systems.

5.1 Boundary conditions and basic design options

5.1.1 Boundary conditions on a storage

Fig.5.1 shows the basic working scheme of heat storage: heat or cold supplied by a heat source is transferred to the heat storage, it is stored in the storage, and later it is transferred to a heat sink to cope with a demand.

Fig. 5.1. Basic working scheme of a storage: heat or cold from a source is transferred to the storage, stored in the storage, and later transferred to a sink.

The problem of storage of heat or cold is inevitably connected with the problem of heat transfer at three points:

- heat transfer source → storage
- heat transfer within the storage
- heat transfer storage → sink

Every application sets a number of boundary conditions originating from the necessary heat transfer:

- temperature: supply temperature ≥ storage temperature ≥ demand temperature
- power
- heat transfer fluid

First, the supply temperature at the source has to be higher or equal to the temperature of the storage, and the storage to the sink. This is necessary to assure that heat will flow in the desired direction no matter how the heat transfer is realized in detail. For cold storage ≥ has to be replaced by ≤. The loading and unloading power, that is the amount of heat transferred in a certain time, is also given as a boundary condition in most applications. In many cases, the application also determines the heat transfer fluid and if it is moved by free or forced convection. With a heat exchanger, it is possible to separate the heat transfer fluid used from source to storage, the storage medium, and the heat transfer fluid used from storage to sink. Such a separation allows to use different media, different pressures, or just different purities like in the case of domestic hot water for drinking. The problem of heat transfer within a PCM and with a boundary for simple geometries was discussed in section 4. Because PCM generally have a low thermal conductivity, the storage design is where the potential to influence heat transfer is the greatest. All applications discussed later reflect this.

5.1.2 Basic design options

The general problem of heat transfer from source to storage and from storage to sink is the same; therefore, the following discussion only treats the last one to avoid repetitions. The heat transfer source → storage and storage → sink can be realized across a heat transfer surface; storage and heat transfer media are in many cases different and in the case of PCM, at least the liquid phase has to be encapsulated. This results in three basic options for heat transfer, shown in fig.5.2:

1. exchanging heat at the surface of the storage,
2. exchanging heat on large surfaces within the storage, and
3. exchanging heat by exchanging the storage medium.

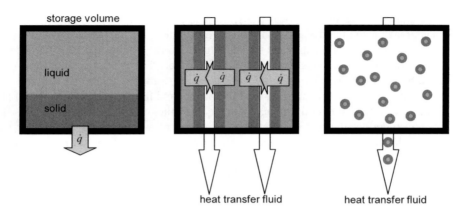

Fig. 5.2. Options to influence heat transfer by storage design: exchanging heat at the surface of the storage (left), on large surfaces within the storage (center), and by exchanging the storage medium (right).

These options are now discussed in more detail.

Option 1: Exchanging heat by heat transfer on the surface of the storage

The left side of fig.5.2 shows a storage with heat transfer on the surface of the storage. Regarding the heat transfer resistance on the surface one can distinguish between

- solid contact to source or sink, or with a fluid which is not moving in a narrow gap; in this case heat is transferred by heat conduction,
- fluid that is moving by free or forced convection, and
- fluid that is evaporated or condensed.

The heat transfer coefficient and thermal resistance for these cases varies considerably. A general discussion of this was done in section 4.1.2., and tab.4.1 shows general properties of heat transfer fluids and PCM layers. If the heat transfer resistance is significant or not depends on the boundary conditions which are the power required by the application and the temperature gradient driving the heat flux, as well as the design of the storage through its geometry like surface to volume ratio. To increase heat transfer one can increase the heat transfer coefficient at the heat transfer surface or increase the area of the heat transfer surface.

Option 2: Exchanging heat by heat transfer on large surfaces within the storage

When the surface on the outside of a storage is not sufficient at a given heat transfer coefficient, one can create additional surface within the storage. Fig.5.2, center, shows a storage with heat transfer on large surfaces within the storage. This approach also includes additional fins, which extend the heat-exchanging surface from the channels for the heat transfer fluid into the storage medium. Larger surfaces within the storage have three beneficial effects. First, the area for heat exchange between the heat transfer fluid and the heat storage medium is increased. Second, the maximum thickness of the PCM layers and thus penetration length of the phase front is significantly reduced; this reduces the overall heat transfer resistance and therefore the temperature drop at the end of a charging or discharging process. The third consequence comes from the fact that in most cases a pump or a fan is used to move the heat transfer fluid; this results in forced convection which increases the heat transfer coefficient in the heat transfer fluid significantly.

Option 3: Exchanging heat by exchanging the storage medium

Fig.5.2 right shows a storage with heat transfer by exchanging the storage medium. In this case, the heat storage medium is also heat transfer medium and heat exchange is by transferring the storage medium. In the case of a hot water heat storage, the option that the heat storage medium and heat transfer medium are the same and that heat exchange is by exchanging the storage medium directly is very common. For example, often a tank contains the hot water and for taking a shower, the stored water is directly discharged. In the case of latent heat storage, the necessity that the heat transfer medium needs to be a fluid requires that the PCM can be pumped in some way. An example is the fluid dispersion of micro encapsulated paraffin, shown in section 2.4.3.2.

The three basic design options have very distinct advantages and disadvantages:

- From top downward, the heat transfer into or out of the storage medium is improved continuously and it gets easier to increase the power of the storage. On the other hand, the fraction of PCM typically decreases from top downwards and thus the storage density also decreases. Power and storage density are in some way competing, but this is not a strict rule.
- In option 1, heat transfer is usually purely passive. In options 2 and 3, heat transfer is usually by actively moving the heat transfer fluid and not by free convection. The consequence is that the storage can be loaded and unloaded when desired. Storages for temperature control are usually constructed using option 1, storages to supply heat or cold by option 2 and 3.

The remaining part of chapter 5 will now discuss how to implement the three options in more detail. For each option, there are several distinct possibilities. For each one, the construction principle is explained, an example is given, and a simple calculation of the heat transfer is derived.

5.2 Overview on storage types

Before starting with a detailed discussion, tab.5.1 gives an overview over the different notations used to describe different storage types. There are three groups of notations. The first one is with respect to the method of heat transfer and used in this book. It has been developed over the past years and is slightly modified to be more specific, compared to earlier publications. The second notation is with respect to the PCM being moved / mixed or not; the storage types are called respectively *dynamic* and *static*. This notation has been used extensively in the past to discuss storage concepts that use artificial mixing to prevent phase separation in PCM that do not melt congruently (section 2.3.1) and / or to improve heat transfer. The third notation is with respect to the heat transfer fluid (HTF) being actively moved or not. In the case of forced convection, the concept is called *active*; for free or no convection the concept is called *passive*. This notation also refers to the possibility of a storage for power regulation using a fan or a pump. Because active and passive types have different heat transfer coefficients in the heat transfer fluid, the third and the first notation have common groups of storage types.

Table 5.1. Different notations to describe storage types.

Notation with respect to the method of heat transfer, used in this book	Notation with respect to PCM being moved / mixed or not	Notation with respect to HTF being actively moved or not
Storages with heat transfer on the storage surface		
Insulated environment	static storage	passive / active
No insulation and good thermal contact between storage and demand	static storage	passive / active
Storages with heat transfer on internal heat transfer surfaces		
Heat exchanger type	static storage	active
Direct contact type	dynamic storage	active
Module type (often called hybrid type when the HTF is a liquid)	static storage	active
Storages with heat transfer by exchanging the heat storage medium		
Slurry type; that is with latent heat contribution	-	active
Sensible liquid type	-	active

142 5 Design of latent heat storages

In the following sections, first storages with heat transfer on the storage surface will be described, and then storages with heat transfer on internal heat transfer surfaces, and finally storages with heat transfer by exchanging the heat storage medium. This order was chosen because it follows the increasing complexity of the design and as a consequence, increasing complexity of the heat transfer calculations.

5.3 Storages with heat transfer on the storage surface

Storages with heat transfer on the storage surface can be described in a simplified manner based on the discussion for the 1-dimensional heat transfer problem in section 4.1.2. They usually have low power, especially when the heat is transferred to a gas where the heat transfer coefficient is relatively small. This can be seen from the long time for the phase front to travel in fig.4.9. A low power restricts the use of the heat stored. Therefore, these storages are applied where the PCM is used for temperature control; the amount of heat stored then determines for how long temperature control can be performed. In cases where good heat transfer is required in the storage material, PCM-graphite materials can be used to improve heat transfer within the storage to assure a constant temperature on the surface as long as possible.

Fig.5.3 shows a storage with heat transfer on the storage surface, which is used for temperature control. The storage buffers losses or gains from the demand (the object which temperature has to be kept constant) to the ambient. The demand can be a human body, temperature sensitive goods, or electronic equipment for example. Often, the demand side is insulated to reduce heat transfer to the environment. Very important, in some cases like the temperature control of the human body, the demand can be a heat source itself.

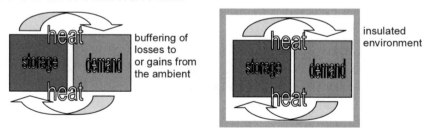

Fig. 5.3. Storage with heat transfer on the storage surface used for temperature control. The storage buffers losses or gains from the demand to the ambient; an insulation helps to prolong the effect.

For storages with heat transfer on the storage surface, the calculation of the heat transfer can be based directly on what was discussed in chapter 4. Some examples are now discussed.

5.3.1 Insulated environment

5.3.1.1 Construction principle and typical performance

The easiest situation to deal with is the situation of temperature control in an insulated environment, as shown in fig.5.4. As an example, the problem that the cool down of an object (heat demand) by heat loss to the environment has to be delayed is discussed. The object itself has only insufficient heat storage capacity, therefore heat losses to the environment are first reduced by an insulation. To extend the effect of temperature control further, a storage is added that slowly supplies heat to equalize the losses through the insulation. Because PCM release the heat at a constant temperature, they are the best choice of storage for this purpose.

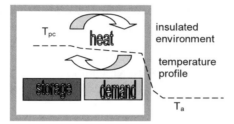

Fig. 5.4. Storage with heat transfer on the storage surface used for temperature control in an insulated environment.

5.3.1.2 Example

As an example, fig.5.5 shows a well-insulated transport box for temperature sensitive goods, which uses macroencapsulated PCM as storage modules.

Fig. 5.5. Transport box from va-q-tec GmbH (picture: va-Q-tec GmbH).

.capsulated PCM modules in this example act as storages with heat the storage surface. Due to the vacuum super insulation, a cooling .ess than 10 W by the PCM is sufficient to sustain -20 °C in the box for 4 . +30 °C environment.

5.3.1.3 Heat transfer calculation

If the demand has only a small thermal mass, the situation is simple. Fig.5.4 shows that the situation is quite similar to the one discussed in section 4.1.2. The boundary effects now include the heat transfer inside and outside the insulated environment and the insulation itself. Looking at typical values for the heat transfer (tab.4.1) it becomes clear that in the calculation of the total thermal resistance the thermal resistance in the PCM can be neglected as a first assumption. This significantly simplifies the problem. The demand temperature will closely follow the storage temperature.

By neglecting the heat storage capacity of the demand, it is possible to make a first estimate. The temperature is controlled until the latent heat ΔH stored in the PCM is lost to the ambient

$$\Delta H = \frac{dQ}{dt} \cdot \Delta t. \tag{5.1}$$

More detailed, the latent heat is $m \cdot \Delta h$, and the overall heat transfer coefficient k_{eff} to the ambient includes the insulation as well as the internal and external boundary layers. Further on, A is the surface of the insulated box. Then

$$\Delta H = m \cdot \Delta_{pc} h = \frac{dQ}{dt} \cdot \Delta t = A \cdot k_{eff} \cdot (T_{pc} - T_a) \cdot \Delta t. \tag{5.2}$$

The solution of the problem is quite simple

$$\Delta t = \frac{m \cdot \Delta_{pc} h}{A \cdot k_{eff} \cdot (T_{pc} - T_a)}. \tag{5.3}$$

It is now easy to estimate what the improvement using the PCM is. The object alone can only release whatever it stores itself by sensible heat within the temperature range ΔT that is tolerated. For the case that the object has a heat capacity of 4 kJ/kgK, and that a temperature range of 5 K is acceptable, the object stores 20 kJ/kg. Assuming that the PCM and the object have the same mass, and that the phase change enthalpy of the PCM is 200 kJ/kg, the ratio of the additional contribution of the PCM to the original heat stored in the object itself is

$$\frac{m \cdot \Delta_{pc} h}{m_{object} \cdot c_{p,object} \cdot \Delta T} = 10. \tag{5.4}$$

The heat stored to stabilize the temperature within the temperature range of 5 K is 10 times larger, and therefore the temperature is also stabilized 10-times longer.

5.3.2 No insulation and good thermal contact between storage and demand

5.3.2.1 Construction principle and typical performance

In this case, the demand is without insulation in contact with the environment, and changes its temperature in an undesired way.

Fig. 5.6. Situation of temperature control with no insulation and good thermal contact between storage and demand.

For temperature control, a storage is installed, as shown in fig.5.6. The storage is in good thermal contact to the demand to ensure sufficient heat transfer.

5.3.2.2 Example

There are many examples for such applications like the use of ice packs on the skin to cool an injury, the use of pocket heaters, or the use of a storage to prevent the overheating of an electronic component.

5.3.2.3 Heat transfer calculation

In this case, the temperature difference that determines the heat transfer is usually not constant. The reason is that there can be significant temperature changes within the demand. Because the temperature difference that determines the heat transfer is usually not constant, heat transfer calculations become much more complex. Numerical models have to be applied to this problem to get reliable and accurate results for the heat transfer and the temperature-time profile in the demand. An exception is the case when the thermal resistance PCM-demand and within the PCM and the demand is much smaller than the one to the ambient.

5.4 Storages with heat transfer on internal heat transfer surfaces

Generally, the heat transfer depends on the heat transfer area, the mode of heat transfer (convection, conduction, radiation) and the corresponding heat transfer coefficient, and the temperature gradient. When trying to improve the heat transfer, it is therefore straightforward to increase the heat transfer area. For storages with heat transfer on the storage surface, as just discussed, this can be done by adding so-called fins to the surfaces of the storage. Usually more efficient is however the use of internal heat transfer surfaces; these are discussed in this section. Storages with heat transfer on internal heat transfer surfaces use a heat transfer fluid to transport heat in and out of the storage. Such storages are able to transfer heat faster to the supply side and / or on the demand side. In contrast to storages with heat transfer on the storage surface, the supply of heat or cold to the demand is usually in the foreground of the application. This means:

- heat is stored for some time, and upon release,
- the match in time or power is in the focus of the application.

There are a number of options to construct storages with heat transfer on internal surfaces. Each option will now be discussed by a description of the construction principle, an example, and a simplified heat transfer calculation as a first overview. The discussion of different storage types focuses on two of the most important storage criteria:

- storage density,
- loading and unloading power.

The following discussion is very simplified to be at the right level for an introduction. In a real setup, the situation is more complex and the stored heat, as well as the power on discharge, will differ somewhat from what is said here. Nevertheless, the simplified discussion is necessary to be able to give an overview on the basic storage types.

5.4.1 Heat exchanger type

Probably the most common kind of storage with heat transfer on internal surfaces is the heat exchanger type. It is derived directly from the construction principle of any kind of heat exchanger to exchange heat between two fluids. To build a storage, on one side of the heat exchanger the fluid is replaced by the PCM.

5.4.1.1 Construction principle and typical performance

As fig.5.7 on the left shows, a heat exchanger type storage consists of a storage vessel that contains the storage medium and an internal heat exchanger. The heat transfer fluid, for example water, flows through the heat exchanger and exchanges heat with the storage medium, here PCM. Usually the heat exchanger consists of a number of pipes distributed equally in the storage volume.

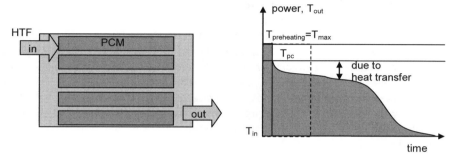

Fig. 5.7. Heat exchanger type: principle (left) and general performance (right). The dashed line indicates the result for a sensible heat storage using the heat transfer fluid only.

Heat exchanger type storages typically have:

- High storage density; up to 95 vol.% are PCM.
- Very high power at the beginning, because the channels of the heat exchanger are filled with some heat transfer fluid that is directly discharged at the beginning.
- A high to medium power later, which depends very much on the proper design of the heat exchanger.

Regarding the storage density, this is easy to understand, however regarding the power it is more difficult. Therefore, the typical features in the power and their origin are now discussed in more detail. The heating or cooling power P of a storage is proportional to the heat capacity of the heat transfer fluid, its volume flow rate, and its temperature difference between inlet and outlet

$$P = \frac{dQ}{dt} = \dot{Q} = c_p \cdot \frac{dV}{dt} \cdot (T_{out} - T_{in}). \tag{5.5}$$

Assuming that the volume flow rate and heat capacity is constant, the power is proportional to the difference between outlet and inlet temperature. In the graph in fig.5.7 on the right side, therefore only one line is used for the schematic representation of the power and the outlet temperature. The integral of the power over time is equal to the heat extracted, and consequently to the heat which was stored. In the graph, this area is shaded in grey. Heat losses are neglected for simplification. For the discussion of the main effects during the unloading of a storage, it is best to use the ideal case that the whole heat storage, which consists of the storage material and heat transfer fluid, initially is preheated to a constant temperature above the phase change temperature. When the unloading of the storage starts, cold heat transfer fluid comes in at the inlet, and heat transfer fluid that was preheated leaves at the outlet. This preheated heat transfer fluid causes a peak in the outlet temperature at the very beginning, as shown in fig.5.7, the temperature of the peak being equal to the preheating temperature. After the preheated heat transfer fluid has left the storage, heat transfer fluid that has entered the storage at the lower inlet temperature T_{in} leaves at the outlet. This heat transfer fluid has only been heated by heat extraction form the storage. The consequence is that the outlet temperature now drops rather fast. As soon as the sensible heat stored above the phase change temperature is extracted from the storage material, the outlet temperature drops below the phase change temperature. The following release of the latent heat slows down the temperature drop to a degree, which is determined by the heat transfer within the storage. There are two main thermal resistances: the thermal resistance in the heat transfer fluid and heat exchanger wall, which is constant, and the thermal resistance in the PCM, which increases with time. Both have been discussed in section 4.1.2 and 4.1.3. In the case of a 2- or 3-dimensional storage however, it is additionally necessary to take into account that the temperature of the heat transfer fluid is not constant in the heat exchanger, but rising from the inlet to the outlet due to the extraction of heat from the storage material. This is why the power and outlet temperature depend very much on the design of the heat exchanger. If the design is in a way that the thermal resistance in the PCM, which increases with time, is small, the constant thermal resistance in the heat transfer fluid and heat exchanger wall dominates and leads to a plateau in the temperature and power as a function of time. Usually the thermal resistance within the PCM, which is increasing with time, has a significant contribution and the power and outlet temperature decrease gradually. The experimental curve in fig.5.14 also shows these general features. Upon heating, the effect is reversed because the outlet temperature is above the phase change temperature. This is sometimes mixed up with subcooling (section 2.1) or hysteresis (section 3.2.2). However, the origins here are the heat transfer in the storage and not a material property. Besides the power and outlet temperature of the heat exchanger type storage, fig.5.7 also includes a dashed line indicating the power and outlet temperature for the sensible heat storage for comparison. At this point, this is only to point out the general differences; the detailed discussion follows in section 5.5.2.

5.4.1.2 Example

Fig.5.8 at the left and center shows two examples of storages where the heat exchanger consists of pipes or flat plate channels. In both examples, a PCM-graphite composite is used to improve the heat transfer from the PCM to the heat exchanger. Another very common method is to use fins attached to the heat exchanger, as shown on the right of fig.5.8.

Fig. 5.8. Left: flat plate heat exchanger (picture: ZAE Bayern). Centre and right: pipe heat exchanger without fins (picture: SGL Technologies) and with fins (picture: ZAE Bayern).

The construction with heat exchangers restricts the flexibility in the design of the storage, but usually the volume fraction of the PCM in the storage is large and its charge / discharge power is comparatively high.

5.4.1.3 Heat transfer calculation

The *total heat stored* Q is the sum of the contributions of the PCM and the heat transfer fluid

$$Q = V^{HTF} \cdot \rho^{HTF} \cdot c_p^{HTF} \cdot \left(T_{\max} - T_{in}\right) \\ + V^{PCM} \cdot \rho^{PCM} \cdot \left(h^{PCM}(T_{\max}) - h^{PCM}(T_{in})\right) \tag{5.6}$$

Usually, the effect of the heat transfer fluid can be neglected because the heat transfer fluid occupies only a small fraction of the storage volume and because it only stores sensible heat. Then

$$Q \cong V^{PCM} \cdot \rho^{PCM} \cdot \left(h^{PCM}(T_{\max}) - h^{PCM}(T_{in})\right). \tag{5.7}$$

It is not possible to calculate the supplied *power* as straight forward as the stored heat. It is necessary to take into account the heat transfer from the PCM to the heat transfer fluid through the heat exchanger before it is discharged from the storage. In section 4, this was discussed on a basic level for different geometries, however assuming that the heat transfer fluid does not change its temperature. This assumption is good when looking at a small portion of a heat exchanger, as shown in fig.4.1, but when looking at the whole heat exchanger the heating of the heat transfer fluid is significant; this is finally, why the storage is used at all. In the following, solutions for the heat transfer will be developed taking into account this effect. The solutions will be based on different restrictions, the most restrictive ones at the beginning leading to the simplest solution.

Analytical model

It might be surprising, but a simple analytical model that can be quite useful can be derived easily. It takes into account the temperature change of the heat transfer fluid, but assumes that on the other side of the heat exchanger the PCM temperature is constant. The starting point is the solution for heat exchange in a heat exchanger. Fig.5.9 shows a heat exchanger that exchanges heat supplied by fluid 2 to heat up fluid 1.

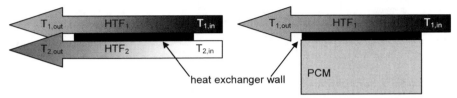

Fig. 5.9. Left: heat exchanger exchanging heat from fluid 2 to fluid 1; dark = low temperature. Right: approximation of a heat storage by a heat exchanger.

For such heat exchangers, the heat flux or power from fluid 2 to fluid 1 is given by

$$\dot{Q} = A \cdot k \cdot \Delta T_{lm} . \tag{5.8}$$

Here A is the heat exchanging area, k is the overall heat transfer coefficient including heat transfer coefficients of the fluids and heat conduction through the heat exchanger material, and ΔT_{lm} is the so-called logarithmic mean temperature difference between fluid 1 and fluid 2 given by

$$\Delta T_{lm} = \frac{\Delta T_{in} - \Delta T_{out}}{\ln \frac{\Delta T_{in}}{\Delta T_{out}}} . \tag{5.9}$$

The heat transferred to fluid 1 will raise the temperature of fluid 1 and

$$\dot{Q} = A \cdot k \cdot \frac{\Delta T_{in} - \Delta T_{out}}{\ln \frac{\Delta T_{in}}{\Delta T_{out}}} . \tag{5.10}$$

If fluid 2 on the left is replaced by a PCM, as shown on the right in fig.5.9, the heat exchanger becomes a heat exchanger type heat storage. Similar to fluid 2 on the left, the PCM will in reality not have a constant temperature along the heat exchanger and the temperature will also change with time. Nevertheless, as a first approach it is useful to assume a constant temperature equal to the phase change temperature along the heat exchanger and constant with time. This will not only give a simple analytical solution, it will also be valid over a wide range as shown later. In a first step, the solution for the outlet temperature and power is now derived; in a second step follows the determination of the conditions when they are valid. Assuming that the PCM is everywhere at the phase change temperature means that all inlet and outlet temperatures of fluid 2 can be replaced by the phase change temperature T_{pc}. Then, the subscript 1 for fluid 1 can be eliminated, as only one fluid is left. Eq.5.5 and eq.5.10 now give

$$c_p \cdot \frac{dV}{dt} \cdot (T_{out} - T_{in}) = A \cdot k \cdot \frac{(T_{pc} - T_{in}) - (T_{pc} - T_{out})}{\ln \frac{T_{pc} - T_{in}}{T_{pc} - T_{out}}} \quad (5.11)$$

$$= A \cdot k \cdot \frac{T_{out} - T_{in}}{\ln \frac{T_{pc} - T_{in}}{T_{pc} - T_{out}}}$$

Hereby c_p is the heat capacity and dV/dt the volume flow rate of fluid 1. To solve this equation for T_{out}, first all parts including T_{out} are arranged on the left side

$$\ln \frac{T_{pc} - T_{in}}{T_{pc} - T_{out}} = \frac{A \cdot k}{c_p \cdot \frac{dV}{dt}} \quad (5.12)$$

Then, the ln is eliminated, which gives

$$T_{out} - T_{in} = (T_{pc} - T_{in}) \cdot \left[1 - \exp\left(-\frac{A \cdot k}{c_p \cdot \frac{dV}{dt}} \right) \right] \quad (5.13)$$

Eq.5.13 is the simple solution to calculate the outlet temperature when the inlet temperature, the parameters of the heat exchanger, and the phase change temperature are known. For large area A and heat transfer coefficient k, and small heat capacity c_p and volume flow dV/dt of the heat transfer fluid, the value of the exponential function becomes zero and the outlet temperature will approach the phase change temperature. In the opposite case, the value of the exponential function becomes one and the outlet temperature remains close to the inlet temperature. This agrees with what is expected from the general performance of heat exchangers.

When a certain outlet temperature is desired, it is possible to derive a general difference between gases and liquids as heat transfer medium, for storages with the same A and dV/dt. Assuming that k is mainly due to α of the heat transfer fluid, the term k / c_p is approximately equal to α / c_p. Gases have a low heat capacity and do not store and transport as much heat as similar volumes of a liquid. A typical ratio of the heat capacity of a liquid to the one of a gas is 10^3. The ratio of the heat transfer coefficients (tab.4.1) is in the order of 10 to 10^2. Therefore, the ratio α / c_p is 10 to 100 times larger for gases than for liquids and as a consequence, it is usually easier to heat a gas than a liquid to a desired temperature.

Using (5.13), the power of the storage is

$$P = \dot{Q} = c_p \cdot \frac{dV}{dt} \cdot (T_{out} - T_{in}) \quad (5.14)$$

$$= c_p \cdot \frac{dV}{dt} \cdot (T_{pc} - T_{in}) \cdot \left[1 - \exp\left(-\frac{A \cdot k}{c_p \cdot \frac{dV}{dt}}\right)\right]$$

Because constant conditions on the PCM side were assumed, there is no time dependency in the solution. Nevertheless, the solution gives important insight into the influence of the main parameters. An important feature of the solution is that it depends only on the total heat transfer area. Consequently, the total power of several storages in parallel is the same, as long as the total heat transfer area A and total volume flow rate dV/dt is the same.

After deriving the solution for the outlet temperature and power, it is necessary to determine the conditions when they are valid. The simple solution was derived using the approximation that the one side of the heat exchanger contains PCM that is always at the phase change temperature. In other words, it is now necessary to find out when this simplification leads to significant errors in the results. The error becomes significant when, upon cooling, the thermal resistance s/λ in the solid PCM-layer leads to a significant temperature drop below T_{pc} at the heat exchanger surface, as shown in fig.5.10.

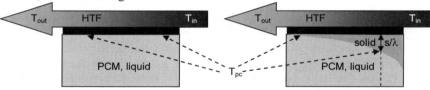

Fig. 5.10. Phase front and thermal resistance s / λ in the solid PCM-layer initially, and at a later time.

To derive a quantitative value, the thermal resistance of the heat exchanger wall is neglected and the problem is treated locally as 1-dimensional, as in section 4.1.2. From eq.4.11 follows that the temperature difference causing the heat transfer is reduced like the ratio of the thermal resistances. If an error of 10 % due to the simplification is acceptable, then the solutions are acceptable as long as

$$\frac{s/\lambda}{1/\alpha + s/\lambda} < 10\%. \quad (5.15)$$

For this 10 % criterion, the maximum distance of the phase front to the heat exchanger surface s_{max} is given in tab.5.2 for typical values of λ and α (tab.4.1). The distance s_{max} is reached first at the inlet.

Table 5.2. Maximum thickness s_{max} of solid PCM layer to have an error smaller 10 % due to the simplification.

	PCM λ = 0.25 W/mK	PCM-graphite matrix λ = 25 W/mK
α = 25 W/(m²K)	s_{max} = 5 mm	s_{max} = 500 mm
α = 250 W/(m²K)	s_{max} = 0.5 mm	s_{max} = 50 mm
α = 2500 W/(m²K)	s_{max} = 0.05 mm	s_{max} = 5 mm

For α = 25 W/m²K, typical for air as heat transfer fluid, the simplification can already be used for the pure PCM up to a PCM layer thickness of 5 mm. This is already typical for many applications. For a PCM-graphite composite, s_{max} is even 500 mm. For α = 250 W/m²K, typical for water as heat transfer fluid, two different cases can be observed. For the pure PCM, the maximum thickness allowed before the simplification leads to serious errors in the result is only 0.5 mm. This is practically never sufficient for real applications. If a PCM-graphite matrix is used, good results can be achieved up to a PCM layer thickness of 50 mm. For α = 2500 W/m²K, the approach is useless for the pure PCM, but applicable for a PCM-graphite composite up to a layer thickness of 5 mm.

Numerical model: 2-dimensional

The analytical model just derived can give valuable information on the influence of some parameters on the power and outlet temperature of the heat transfer fluid. However, because of the assumption that the temperature on the PCM side is everywhere and always equal to the phase change temperature T_{pc}, a time dependency of the power and outlet temperature cannot be derived with this model. Further on the inlet temperature also has to be constant in the analytical model. To get a description of the changes with time it is necessary to take into account the moving phase boundary, which requires a 2-dimensional model. Probably the easiest way to do this is to extend the numerical model described in section 4.2.1 to two dimensions.

To show how a 2-dimensional numerical model works the storage shown in fig.4.1 is discussed as example. The storage consists of four identical storage modules, separated by an insulating layer. It is therefore sufficient to develop a model for a single storage module. Each storage module consists of a heat exchanger that transfers heat from the PCM on its left and right side to the heat transfer fluid. Fig.5.11 shows the 2-dimensional net for the numerical description of a single storage module. It includes nodes for the PCM, the heat transfer fluid, and the ambient.

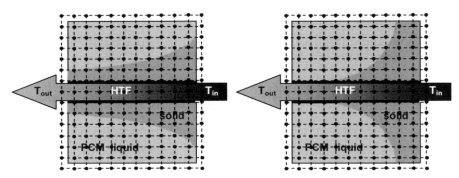

Fig. 5.11. Sketch of a 2-dimensional numerical model including nodes for the PCM, the heat transfer fluid, and the ambient. Different thermal resistances lead to different shapes of the phase front.

The storage in fig.4.1 uses a PCM-graphite matrix as storage material. This storage material is very suitable for this example because there is only heat conduction in the storage material and no convection, exactly as in the numerical model. The movement of the heat transfer fluid and the associated heat transfer by forced convection can be treated in a simple manner: each time step the heat transfer fluid in a node within the heat exchanger, more specifically its value of the temperature, is moved from one node to the next one. Because of this approach, the time step in the simulation must be adjusted to give the correct volume flow. This simplified treatment is possible because the heat transfer fluid is moved by forced convection caused by the pump and not by a density gradient connected to a temperature profile.

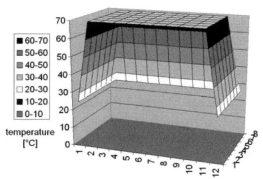

Fig. 5.12. Numerical model of the storage with initial temperatures.

Fig.5.12 shows the nodes with the initial temperatures. Because of the symmetry of the storage module, only one side of the storage module is shown. The storage has a length of 1 m represented by 10 nodes (nodes 2 to 11), a height of 0.1 m, and a depth of 0.05 m represented by 6 nodes (nodes 2 to 7). The PCM-graphite matrix is oriented in a way that heat transfer is preferably towards the heat exchanger and limited by air gaps perpendicular to the heat exchanger. The ambient, represented by the outer row of nodes in fig.5.12, has a constant temperature of 25 °C. The melting temperature of the PCM is 55 °C, and it is initially at a temperature of 70 °C. The water in the heat exchanger, represented by the nodes in the front line, is initially also preheated to 70 °C. Then, the storage is cooled down with water at an inlet temperature of 10 °C flowing at a rate of 1 liter/min form the right to the left. Fig.5.13 shows the result of the simulation in a series of pictures.

5 Design of latent heat storages

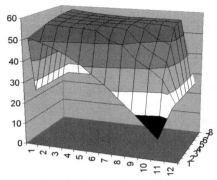

After 2 min: water with 10 °C cools the inlet; most of the PCM is at T_{PC}; water leaves at 51 °C.

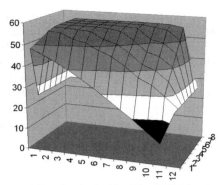

After 4 min: half of the PCM is at T_{PC}; water leaves at 48 °C.

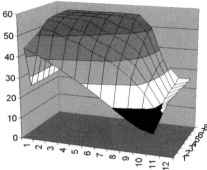

After 6 min: less than half of the PCM is at T_{PC}; water leaves at 45 °C.

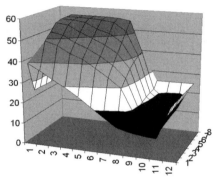

After 8 min: most of the PCM is below T_{PC}; water leaves at 40 °C.

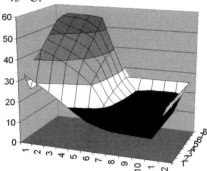

After 10 min: most of the PCM is below T_{PC}; water leaves at 34 °C.

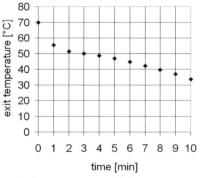

Exit temperature of the water as a function of time.

Fig. 5.13. Result of the numerical simulation for the cooling of one storage module.

The last picture in fig.5.13 shows the typical features already shown in fig.5.7. The exit temperature is equal to the initial temperature of 70 °C for a short time. Then the temperature drops and builds a more or less pronounced plateau at a temperature below T_{pc}. How much below depends on several parameters, most important the heat transfer resistances. In the example, due to the good heat transfer towards the heat exchanger by the graphite matrix, the storage material is first cooled at the inlet of the storage module and then successively towards the rear end. This is often compared to a candle burning down from one end to the other. The typical temperature profile is shown on the right in fig.5.11. On the left, is the typical temperature profile for a bad heat transfer towards the heat exchanger, which is typical for pure PCM due to their low thermal conductivity. In that case, the phase front moves away from the heat exchanger everywhere at a similar speed. The heat exchanger can then be used as a symmetry axis, for example for cylindrical coordinates when the heat exchanger is built from pipes.

As with 1-dimensional models, it is very important to check also the accuracy of 2- and 3-dimensional numerical models by a comparison between numerical and experimental results. For the storage just discussed, fig.5.14 shows such a comparison between the outlet temperature history recorded in an experiment and results from a numerical simulation. The agreement between measurement and simulation is very good.

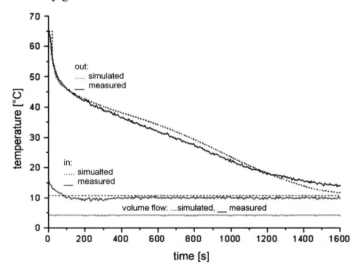

Fig. 5.14. Comparison of the outlet temperature history recorded in an experiment and from a simulation (picture: ZAE Bayern).

At this point, it is necessary to make a very important comment on the amount of heat stored and the storage density. At the beginning of section 5.4.1.3 the total heat stored is calculated assuming that the heat storage is initially loaded / preheated to a maximum temperature and completely unloaded / cooled down to the inlet temperature of the heat transfer fluid. This is a very idealistic assumption, but necessary to simplify the discussion. In reality, this assumption is usually not fulfilled. If in fig.5.14 for example, a temperature of 30 °C is needed at the outlet, only a part of the total stored heat is also useful. The *useful stored heat*, that is the heat that can be taken from the storage at a useful temperature, can be very different from the total stored heat and depends on many parameters. The same difference has to be made for the storage density of course; only a part of it is useful.

5.4.1.4 Further information

The models discussed here to describe the heat transfer in a storage were purely analytical or purely numerical. It is however also possible to combine analytical and numerical approaches, as done by Dolado et al. 2006. They describe a model that treats the movement of the heat transfer fluid by a numerical approach, with nodes as shown in fig.5.11, while the heat transfer from the PCM to the heat exchanger is described by the analytical solution in eq.4.11. The latter includes the moving of the phase front and the heat transfer from the heat exchanger to the heat transfer fluid. This significantly simplifies the numerical problem and saves computation time; the drawback is that the model is more restrictive. For the case of a storage to be integrated in an air conditioning system, Dolado et al. 2006 compare this model with three other models with respect to accuracy and advantages.

An important issue that has not been covered here is the extension of the heat exchanger surface using fins, as shown in fig.5.8 on the right. The inclusion of fins in analytical or numerical heat transfer calculations significantly complicates the discussion. It is treated in many textbooks and is no specialty of PCM. Information on the modeling of fins by analytical and approximate models can be found in Dincer and Rosen 2002, Lamberg and Siren 2003a, Lamberg and Siren 2003b, and Lamberg 2004. A review of such literature can also be found in Zalba et al. 2003, and Kenisarin and Mahkamov 2007.

5.4.2 Direct contact type

Storages of the heat exchanger type can give high loading and unloading power if the heat transfer within the storage material is excellent. To achieve excellent heat transfer within the PCM, graphite or metallic structures can be combined with the PCM, or fins can be attached to the heat exchanger. These measures are however at additional cost. Further on, they are not always an improvement. While these structures generally improve the heat transfer in the solid state of the PCM, they can have a negative effect in the liquid phase if they suppress convection. With convection, especially with forced convection, it is possible that the heat transfer is already sufficient. The direct contact concept takes advantage of convection and actually enforces it. In addition, it can also solve the problem of phase separation in PCM that do not melt congruently (section 2.3.1).

5.4.2.1 Construction principle and typical performance

In a direct contact storage the heat transfer fluid is in direct contact with the PCM; this eliminates the thermal resistance of the heat exchanger wall. To be able to separate PCM and heat transfer fluid in order to supply the heat transfer fluid to the sink, it is necessary that heat transfer fluid and PCM do not mix with each other. When the PCM is a salt hydrate, oil can be used as heat transfer fluid; when the PCM is a paraffin, water can be used. In a direct contact storage, the storage vessel is filled up to about 90 vol.% with PCM. When the heat transfer fluid has a lower density than the PCM, the heat transfer fluid is pumped into the bottom of the storage, as shown in fig.5.15. The heat transfer fluid is released as droplets, which, due to their lower density, rise from the bottom. On their way upwards, the droplets exchange heat with the PCM by direct contact and melt or solidify the PCM. When the droplets reach the surface of the PCM, they are collected and the circuit of the heat transfer fluid is closed.

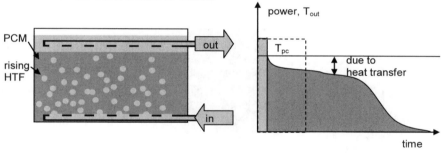

Fig. 5.15. Direct contact heat storage: principle and general performance.

An additional advantage of direct contact systems is that the droplets of the heat transfer fluid cause convection and mixing within the PCM. Direct contact storages therefore also belong to the category of dynamic storages (tab.5.1).

The case that the heat transfer fluid has a lower density than the PCM, as shown in fig.5.15, applies to the use of salt hydrates as PCM and oil as heat transfer fluid. For paraffin as PCM and water as heat transfer fluid, the water has to be supplied at the top of the storage, settles down due to its higher density compared to the paraffin, and is collected at the bottom. In this case, the density difference is however smaller than for the combination of salt hydrate and oil.

Direct contact storages have similar performance to heat exchanger type storages. Typically, direct contact type storages have:

- high storage density: because up to 90 vol. % are PCM

- high to medium power, because the heat transfer between storage medium and heat transfer medium is improved by direct contact and by forced convection
- an extra pump and heat exchanger to separate the heat transfer fluid circulating in the storage, which might be contaminated with PCM, from the demand circuit.

Because of the mixing effect, direct contact systems have been investigated intensively for salt hydrates that do not melt congruently like $CaCl_2 \cdot 6H_2O$, $Na_2SO_4 \cdot 10H_2O$, $Na_2HPO_4 \cdot 12H_2O$, $NaCO_3 \cdot 10H_2O$, $Na_2S_2O_4 \cdot 5H_2O$ (Lindner 1984, Fouda et al. 1984, Farid and Yacoub 1989, Farid et al. 2004). Often, extra water has been added to the salt hydrate to prevent clogging. He et al. 2004 also published a detailed description of an experimental setup for a direct contact system using paraffin RT5 as PCM for cooling.

5.4.2.2 Example

Fig.5.16 shows an experimental setup of a direct contact system at the ZAE Bayern. The PCM investigated was $CaCl_2 \cdot 6H_2O$ and a special oil was used as heat transfer fluid. To be able to distinguish visually between oil and liquid PCM, the oil was colored red.

Fig. 5.16. Experimental setup of a direct contact system and PCM-oil mixture in the solidified state (pictures: ZAE Bayern)

The oil is pumped into the storage at the bottom, and rises due to its lower density to the top of the storage. There, it forms a layer and is then pumped back into the circuit. When the PCM solidifies the rising of the oil is slowed down and finally oil can be trapped within the solid PCM (fig.5.16, left).

5.4.2.3 Heat transfer calculation

The heat transfer in a direct contact system depends on many parameters, the most important ones are the viscosity and the density of the PCM and the heat transfer fluid as a function of temperature, the flow rate and bubble size of the heat transfer fluid. The bubble size further on depends on the surface energy, a parameter that is dependent on the combination of PCM and heat transfer fluid. The large number of parameters and the complexity of the theoretical background make the description of the heat transfer a difficult task. To the authors knowledge there is currently no simple mathematical model available to calculate the heat transfer in a direct contact system, not even an empirical model.

5.4.2.4 Further information

He et al. 2004 published a detailed description of an experimental setup for a direct contact system using paraffin RT5 as PCM and water as heat transfer fluid. The description includes many pictures and a set of experimental data. Lindner 1984, Fouda et al. 1984, Farid and Yacoub 1989, and Farid et al. 2004 have described systems using salt hydrates as PCM in detail.

A special kind of direct contact type with solid-liquid phase change is the so-called Galisol principle.

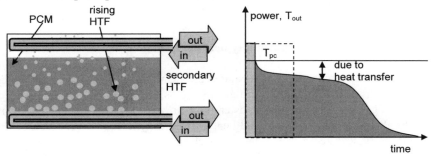

Fig. 5.17. Galisol type: principle and general performance.

As fig.5.17 shows, it has two distinct differences compared to conventional direct contact storages. First, the heat transfer fluid used within the storage to exchange heat with the PCM remains within the storage and exchanges heat to a secondary heat transfer fluid outside the storage. This is done via heat exchangers located inside the storage; for heat input at the bottom and for heat extraction at the top of the storage. Second, the primary heat transfer fluid also changes phase, from liquid to vapor. To make the heat transfer fluid rise from the bottom after it is heated, it is vaporized. The vapor has a very low density compared to the PCM, rises fast to the top, and releases heat at the heat exchanger at the top by condensing. After the heat transfer fluid is condensed to a liquid, it sinks down again. For this, it is necessary that the liquid heat transfer fluid has a higher density than the PCM. The advantages of the system are the large density difference between the liquid and the vaporized heat transfer fluid regarding the dynamics of the storage, and the vaporization and condensation at the two heat exchangers, which leads to large heat transfer coefficients. The disadvantage is that it is hard to find suitable materials as heat transfer fluid. In the past, often CFCs (chlorofluorocarbons) were used which are today off the market because of their potential to destroy the ozone layer. Some information on this special type of storage is available in the publication by Naumann et al. 1989.

5.4.3 Module type

The third kind of storages with heat transfer on internal heat transfer surfaces is the module type storage. It uses macroencapsulated PCM modules for heat storage. Using macroencapsulated PCM modules is a big advantage regarding manufacturing and marketing (Mehling 2001). The manufacturer can solve problems regarding the function of a PCM, like volume change and heat exchange by the geometry, within the PCM module. The macroencapsulated PCM module can then be a product by itself with its own warranty. Any customer buying the modules therefore does not have to bother with problems related to PCM. Further on, the design of module type storages is extremely flexible.

5.4.3.1 Construction principle and typical performance

Instead of a heat exchanger that separates the PCM from the heat transfer fluid, the PCM is macroencapsulated. As fig.5.18 shows, the PCM modules are placed in the storage tank and the heat transfer fluid flows in the tank exchanging heat with the PCM modules. Module and heat exchanger type can be very similar regarding the typical geometries. The approach that the modules contain the PCM however allows a flexible construction of the storage and a prefabrication of the modules suitable for series production.

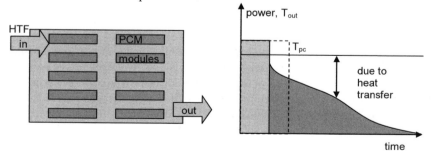

Fig. 5.18. Module type: principle and general performance.

The PCM-modules are usually macroencapsulated PCM in form of bags, dimple sheets, flat plates, and spherical capsules (section 2.4.3.1). Spherical capsules, often called nodules, are the most common encapsulation for this type of storage. One reason is that spherical modules arrange themselves automatically when filled into a tank. The highest possible package density (PCM fraction of the volume) is 74 vol.% for spheres of the same size. This is why module type storages can also be called *hybrid storages* (tab.5.1), when the heat transfer fluid is a liquid; they combine latent heat storage with a considerable fraction of sensible heat storage due to the amount of heat transfer fluid in the storage tank of more than 25 vol.%. Module type storages are mostly used with liquids as heat transfer fluid, like water, water-salt or water-glycol mixtures, but have also been investigated with air. With air as heat transfer fluid however usually flat plates, bags or dimple sheets are preferred because of the larger surface to volume ratio (discussion in section 4.1).

Module type storages (fig.5.18) typically have:

- medium storage density, due to the combination of latent heat storage with sensible heat storage
- high power at the start, due to the amount of preheated heat transfer fluid within the storage
- lower power later, due to the heat transfer between PCM-modules and the heat transfer fluid.

It is important to stress here again that this discussion is of a qualitative nature. The statements regarding the general performance of a storage type simplify the discussion in this introductory text. The wide variety of geometries and sizes of the modules, the volume flow of the heat transfer fluid, and the boundary conditions given by an application usually make things more complicated. For example, it is possible to achieve the same geometry with modules as with flat plate heat exchangers. Then, the heat transfer and power will be the same. However, usually this is not done because the production methods for modules and heat exchangers prefer different geometries and dimensions.

5.4.3.2 Examples

There are many examples of module type storages. The most common ones use modules with a spherical shape or flat plates. Fig.5.19 shows two laboratory set-ups for air as heat transfer fluid, where the construction and the modules are visible.

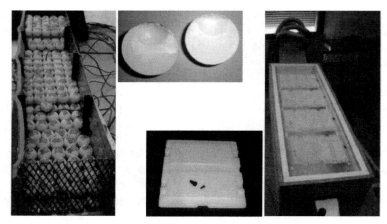

Fig. 5.19. Laboratory setup of a storage with air as heat transfer fluid using spherical capsules on the left (pictures: University of Ljubljana) and flat plates on the right (pictures: ZAE Bayern)

Both laboratory setups were built and tested for applications in solar assisted building ventilation. The left, described by Arkar and Medved 2002, uses spherical capsules filled with a PCM. The right side shows a setup with flat plates, investigated at the ZAE Bayern.

5.4.3.3 Heat transfer calculation

The *heat stored* is simply calculated from the sensible components of the heat transfer fluid and the PCM plus the latent heat of the PCM assuming that isothermal conditions hold at the start and the end. Then

$$Q = V^{HTF} \cdot \rho^{HTF} \cdot c_p^{HTF} \cdot (T_{max} - T_{in}) \\ + V^{PCM} \cdot \rho^{PCM} \cdot \left(h^{PCM}(T_{max}) - h^{PCM}(T_{in})\right) \tag{5.16}$$

As temperature difference, the maximum temperature that the storage is heated to, T_{max}, and the minimum temperature it can be cooled down to, T_{in}, are used. When air is the heat transfer fluid, its contribution to the stored heat can be neglected because of its low heat capacity.

The calculation of the power supplied to or by the storage is more complicated to calculate. This is because it is necessary to take into account heat transfer within the PCM modules, and from the PCM-modules to the heat transfer fluid before it is leaves the storage. Often, a porous medium model is used to describe module type storages. These models treat the storage as a packed bed of PCM modules. A short derivation and description of the porous medium approach is now described. Fig.5.20 shows the heat transfer model for a packed bed. The storage is divided into slabs (volume elements), each represented by a node denoted by the variable i. Each slab includes several PCM modules. The main assumption is that all PCM modules in a slab behave the same, and therefore can be treated as one source / sink mathematically. This is possible because also all the heat transfer fluid in the slab is treated the same way and is represented by a single node.

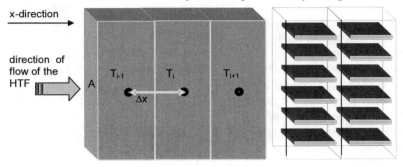

Fig. 5.20. Heat transfer model for a packed bed. Left: division of the storage into slabs (volume elements) represented by a node denoted by the variable i. Right: two slabs with flat PCM modules.

The heat gain $\Delta Q_{i,j}$ of the heat transfer fluid represented by node i during time step j leads to a change of the temperature of that node from time step j to j+1. Because the heat transfer fluid stores only sensible heat, the heat gain and temperature change are related by

$$\Delta Q_{i,j} = A \cdot \Delta x \cdot \rho \cdot c_p \cdot (T_{i,j+1} - T_{i,j}). \tag{5.17}$$

To calculate the heat gain $\Delta Q_{i,j}$ in node i several terms of heat transfer have to be taken into account. These are now derived separately.

Heat gain $\Delta Q_{i,j}$ in node i due to heat conduction:

There are two contributions of heat gain due to heat conduction from the two neighbor slabs

$$\begin{aligned}\Delta Q_{i,j} &= -\Delta t \cdot A \cdot \lambda \cdot \frac{T_{i,j} - T_{i-1,j}}{x_i - x_{i-1}} + \Delta t \cdot A \cdot \lambda \cdot \frac{T_{i+1,j} - T_{i,j}}{x_{i+1} - x_i} \\ &= \Delta t \cdot A \cdot \lambda \cdot \left(\frac{T_{i+1,j} - T_{i,j}}{x_{i+1} - x_i} - \frac{T_{i,j} - T_{i-1,j}}{x_i - x_{i-1}} \right) \end{aligned} \tag{5.18}$$

Heat gain $\Delta Q_{i,j}$ in node i due to heat transfer associated with the flow by pumping the heat transfer fluid:

In contrast to a general treatment of heat transfer by convection, this special case is not too complicated to simulate. The reason is that the convection is enforced by the pump and not due to density gradients caused by temperature differences. As a further simplification, the heat transfer fluid is assumed to be incompressible. To take into account heat gains or losses by a movement of the heat transfer fluid, it is necessary to know the amount of heat transfer fluid that enters or leaves the slab i and how much heat the heat transfer fluid carries in or out. The volume that enters or leaves in a time interval Δt is $\Delta V = \Delta t \cdot A \cdot \Delta x / \Delta t$. The heat transfer fluid flows into x-direction, its speed denoted by u. Then, heat is transferred by heat transfer fluid entering from node i-1 at a temperature $T_{i-1,j}$, and by fluid leaving from node i at a temperature $T_{i,j}$

$$\Delta Q_{i,j} = \Delta t \cdot A \cdot u \cdot \rho \cdot c_p \cdot (T_{i-1,j} - T_{i,j}). \tag{5.19}$$

The heat gain is positive when $T_{i-1,j} > T_{i,j}$, that is when the fluid entering slab i has a higher temperature than the one leaving it.

Heat gain due to heat exchange with the PCM modules in a slab (fig.5.20, right):

The main assumption now is that all PCM modules in a slab behave the same and are treated as one source / sink mathematically. This is possible because also all the heat transfer fluid in the slab is treated the same and represented by a single node. The surface area of the PCM modules in a volume element is the total surface area of all PCM modules A^{PCM} multiplied by the fraction for a volume element $A \cdot \Delta x / V$. Then, the heat gain from the PCM modules by heat transfer from the surface of all the PCM modules in the slab, at a surface temperature T^{PCM} is

$$\Delta Q_{i,j} = \Delta t \cdot A^{PCM} \cdot A \cdot \Delta x \cdot \frac{1}{V} \cdot \alpha \cdot (T_{i,j}^{PCM} - T_{i,j}). \tag{5.20}$$

Taking all parts together and dividing by $\Delta t \cdot A \cdot \Delta x$ gives the heat gain per time interval of a node (volume element)

$$\frac{\Delta Q_{i,j}}{\Delta t \cdot A \cdot \Delta x} = \rho \cdot c_p \cdot \frac{T_{i,j+1} - T_{i,j}}{\Delta t}$$

$$= \lambda \cdot \left(\frac{T_{i+1,j} - T_{i,j}}{x_{i+1} - x_i} - \frac{T_{i,j} - T_{i-1,j}}{x_i - x_{i-1}} \right) / \Delta x$$

$$- \rho \cdot c_p \cdot u \cdot \frac{T_{i,j} - T_{i-1,j}}{\Delta x}$$

$$+ \frac{A^{PCM}}{V} \cdot \alpha \cdot \left(T_{i,j}^{PCM} - T_{i,j} \right) \tag{5.21}$$

In differential form this is

$$\rho \cdot c_p \cdot \frac{\partial T}{\partial t} = \lambda \cdot \frac{\partial^2 T}{\partial x^2} - \rho \cdot c_p \cdot u \cdot \frac{\partial T}{\partial x} + \frac{A^{PCM}}{V} \cdot \alpha \cdot \left(T^{PCM} - T \right). \tag{5.22}$$

For the complete description, it is necessary to take into account that the heat transfer fluid fills only a fraction f of the volume; the PCM modules fill the remaining volume fraction. Then eq.5.21 becomes

$$\frac{\Delta Q_{i,j}}{\Delta t \cdot A \cdot \Delta x} = f \cdot \rho \cdot c_p \cdot \frac{T_{i,j+1} - T_{i,j}}{\Delta t} \tag{5.23}$$

$$= \lambda \cdot \left(\frac{T_{i+1,j} - T_{i,j}}{x_{i+1} - x_i} - \frac{T_{i,j} - T_{i-1,j}}{x_i - x_{i-1}} \right) / \Delta x$$

$$- f \cdot \rho \cdot c_p \cdot u \cdot \frac{T_{i,j} - T_{i-1,j}}{\Delta x}$$

$$+ \frac{A^{PCM}}{V} \cdot \alpha \cdot \left(T_{i,j}^{PCM} - T_{i,j} \right)$$

and eq.5.22 becomes

$$f \cdot \rho \cdot c_p \cdot \frac{\partial T}{\partial t}$$

$$= \lambda \cdot \frac{\partial^2 T}{\partial x^2} - f \cdot \rho \cdot c_p \cdot u \cdot \frac{\partial T}{\partial x} + \frac{A^{PCM}}{V} \cdot \alpha \cdot \left(T^{PCM} - T \right) \tag{5.24}$$

Eq.5.23 and eq.5.24 describe the time evolution of the temperature of the heat transfer fluid, knowing the surface temperature of the PCM modules T^{PCM}. The calculation of the temperature at the surface of the PCM modules T^{PCM} is remaining. In an accurate solution, the geometry of the modules and the movement of the phase boundary are taken into account. The modules in different slabs will then be in different states at a fixed time and have to be treated separately. In a more crude solution, the temperature T^{PCM} is again kept constant and set equal to the phase change temperature. For this case, the solution can be simplified further because modules in different slabs will still have the same surface temperature. The model then is similar to the one for the heat exchanger type in fig.5.9, just with a different arrangement of the surface area of the modules. As eq.5.13 and eq.5.14 showed, this has no influence on the solution, so both solutions should become the same under these conditions. To prove this mathematically, in eq.5.24 T^{PCM} must be set constant, as well as the velocity u of the heat transfer fluid. Further on, the conduction term must be neglected by setting λ to zero. Last but not least, under these assumptions there is no change with time. The solution of eq.5.24 under these conditions then becomes eq.5.13 again.

5.4.3.4 Further information

More information on the porous medium approach can be found in the literature. For example, Ismail and Stuginsky 1999 present a study of possible packed bed models for LHS. A description of a model can also be found in Dincer and Rosen 2002. Pluta 2003, Arkar and Medved 2002, and Arkar and Medved 2005, have presented numerical models for systems with air as heat transfer medium and PCM in spherical capsules. The last reference also describes a comparison with experimental data. Regarding systems with water as heat transfer medium, Wang et al. 2003 compared numerical simulations using the enthalpy method and the heat capacity method for a storage tank packed with spherical and cylindrical PCM modules. Egolf and Manz 1995 performed experimental and numerical investigations for a vertical storage filled with spheres and water as heat transfer fluid. Bédécarrats et al. 1996 describe experiments on a tank filled with nodules and water as PCM. The data are compared with simulations using a packed bed approach where during the cooling of the PCM a statistical approach determines the number of nodules showing crystallization. In Bédécarrats and Dumas 1997, they describe more details of the inclusion of subcooling. Heinz and Streicher 2006 present experimental data where the performance of modules filled with a paraffin, a salt hydrate, and a salt hydrate - graphite composite are compared. Recently, Heinz 2007 developed a model that includes also subcooling and implemented it in the commercial building simulation software TRNSYS. These models however do not include the crystallization speed, as described in section 4.2.2.

5.5 Storages with heat transfer by exchanging the heat storage medium

In a module type storage, the main part of the heat is stored in the PCM modules. They remain within the storage vessel and exchange heat with the supply or demand via the heat transfer fluid. The heat transfer to the heat transfer fluid however takes time and has limitations. An increase in the heat exchanging area, more specific an increase in the surface to volume ratio of the modules, will strongly improve heat transfer. When thinking about smaller and smaller modules however, there is a point where the size of the modules is so small that it is possible that they leave the storage and flow with the heat transfer fluid. Then, heat storage medium and heat transfer medium become one fluid which consists of a component that stores heat by a phase change and a component that is always liquid and assures the fluid properties. The limiting case of zero PCM fraction is the sensible liquid storage type.

5.5.1 Slurry type

A liquid, which consist of a component that stores heat by a phase change and a component that is always liquid and assures the fluid properties, is named slurry, *phase change slurry (PCS)*, or *phase change liquid (PCL)*. An example is a mixture of water and microencapsulated PCM. The PCM in the microcapsules stores the heat and the water assures the fluid behavior even when the PCM in the microcapsules is solid. Because a slurry is liquid at all times it can be stored in a conventional tank, discharged directly, pumped through pipes, and heated or cooled by a heat exchanger. Because of the phase change, a phase change slurry has a higher storage density when stored in a storage tank than the pure heat transfer fluid. However, there are additionally several secondary advantages. Due to the phase change, the slurry will transport more heat or cold at the same volume flow rate and the pipe system already acts as a storage. Further on the heat transfer coefficients are improved, which allows the use of smaller heat exchangers etc. Because of these secondary advantages, the same performance can be achieved with smaller systems, or the performance of a system can be enhanced when a phase change slurry is used instead of a conventional heat transfer fluid without a PCM fraction. This is an interesting option for the retrofit of existing systems.

5.5.1.1 Construction principle and typical performance

The construction principle of a slurry type heat storage is quite simple, as a consequence of the properties of a slurry: it is just a storage vessel containing the slurry, as shown in fig.5.21. Actually, also the pipe system now acts as a storage vessel, a fact that is often forgotten but can be significant.

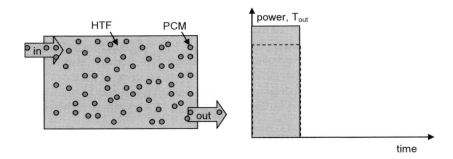

Fig. 5.21. Slurry type: principle and general performance. The dashed line indicates the result for a sensible heat storage using the heat transfer fluid only. The temperature of the slurry is the same as for the sensible heat storage, however due to the presence of the PCM the power is higher.

Slurry type storages typically have:

- increased storage density compared to the pure heat transfer fluid (dashed line in fig.5.21), but typically less than the other types mentioned before
- high power, because more heat is transported per volume of heat transfer fluid, and
- improved heat transfer, as the improved storage density also raises the heat transfer coefficient

It is important to say that in fig.5.21 there is a difference to the discussion of the other storage types. For the slurry, the power and the temperature as a function of time are not represented by the same curve. While the temperature of the slurry is again the same as for the sensible heat storage, the power is higher due to the presence of the PCM. The same applies to the heat stored, which is equal to the integral of the power over time.

5.5.1.2 Example

There are several general kinds of phase change slurries with distinct properties.

An *ice slurry* is water with a fraction of ice particles, as shown at the top in fig.5.22. The water assures that the ice slurry is a fluid; the ice enhances the cold stored and transported with the ice slurry compared to just cold water. A specialty of ice slurries is that they consist of only one component; therefore, the particles disappear completely in the melting process and have to be created again by an ice generator. Often an additive, usually an alcohol or salt, is added to lower the melting temperature of the water; in this case, the slurry is called *binary ice*. The additive assures that when the slurry is cooled slightly below 0 °C, only a fraction of the water forms ice while the additive concentration in the remaining water increases and prevents freezing.

5.5 Storages with heat transfer by exchanging the heat storage medium 171

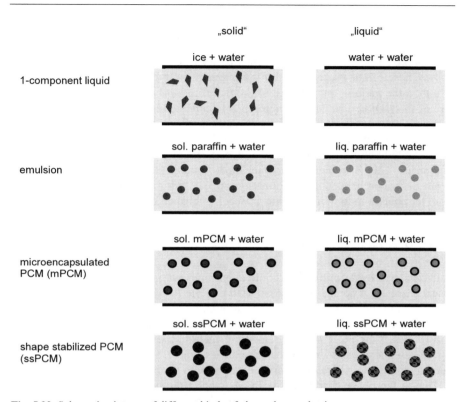

Fig. 5.22. Schematic pictures of different kinds of phase change slurries.

In all other cases, the PCM slurry consists of two components: one that changes phase and another that assures the fluid properties.

A homogeneous mixture of both components is called *emulsion slurry*. This approach is often used when water as carrier fluid carries a PCM that is not soluble in water. For example, paraffin can be mixed with water and stirred until an emulsion is achieved, as shown in fig.5.22. However, such an emulsion is usually not stable for a long time, the paraffin droplets will form larger drops, and eventually a paraffin layer will float on top of a water layer, due to the density difference. The emulsion can however be stabilized by an emulsifier. The molecules of an emulsifier have a hydrophobic and a hydrophilic end and build a stabilizing layer between the hydrophobic paraffin and the water by orienting the hydrophilic group towards the outside where the water is and the hydrophobic group towards the inside where the paraffin is.

An emulsion, even when stabilized, will eventually degrade with time into two separate fluids. To avoid this, the PCM droplets can be microencapsulated, that means a capsule wall protects each droplet, as fig.5.22 shows. An example of a *microencapsulated PCM slurry* shows fig.2.32. The encapsulation prevents the leakage of PCM in the liquid phase, which could otherwise solidify in pipes and block them. It is important that the capsules are sufficiently resistant to the shear stresses occurring in the pumps.

A *shape-stabilized PCM slurry* is based on shape-stabilized PCM (ssPCM) particles. These can be prepared by infiltrating paraffin in HDPE at a higher temperature than the melting temperature of the paraffin, but below the melting temperature of the HDPE. The paraffin is trapped in cells within the HDPE structure; this acts as a supporting structure for the shape and prevents also leakage (section 2.3.4.1). A schematic picture is shown at the bottom of fig.5.22.

Inaba 2000 and Egolf et al. 2003a give excellent overviews on different kinds of slurries. They also mention another kind of slurry made with clathrates, which is not described here. Some information on tests with a clathrate slurry prepared from tetrabutylammonium bromide has been published by Xiao et al. 2006.

5.5.1.3 Heat transfer calculation

A PCM slurry by definition is a liquid at all times with a fraction that undergoes a phase change. Of course, it is possible to call the PCM slurry itself now the heat transfer fluid, but to be consistent and to avoid confusion with the notation used until now this will not be done here. The fraction that undergoes the phase change will be called the PCM and the remaining fraction the heat transfer fluid. This is correct for most cases except the ice slurries where both fractions are from the same material and where the fraction changes with temperature.

Due to the phase change of the PCM fraction the heat storage capacity of the PCM slurry is larger compared to common heat transfer fluids without phase change. The *heat stored* is the sum of the sensible heat of the heat transfer medium and PCM, and the latent heat of the PCM

$$Q = V^{HTF} \cdot \rho^{HTF} \cdot c_p^{HTF} \cdot (T_{max} - T_{in})$$
$$+ V^{PCM} \cdot \rho^{PCM} \cdot (h^{PCM}(T_{max}) - h^{PCM}(T_{in})) \quad (5.25)$$

Using f as the volume fraction of PCM this can be rewritten to

$$Q = V \cdot \begin{bmatrix} (1-f) \cdot \rho^{HTF} \cdot c_p^{HTF} \cdot (T_{max} - T_{in}) \\ + f \cdot \rho^{PCM} \cdot (h^{PCM}(T_{max}) - h^{PCM}(T_{in})) \end{bmatrix} \quad (5.26)$$

The *power supplied* is equal to the heat transported in a time interval

$$P = \frac{d}{dt} Q$$
$$= \frac{dV}{dt} \begin{bmatrix} (1-f) \cdot \rho^{HTF} \cdot c_p^{HTF} \cdot (T_{max} - T_{in}) \\ + f \cdot \rho^{PCM} \cdot (h^{PCM}(T_{max}) - h^{PCM}(T_{in})) \end{bmatrix} \quad (5.27)$$

For 1-component slurries, the first term vanishes and $f = 1$.

A very common example for the use of a PCM slurry is the use of ice slurries in cooling applications. A cold-water storage with a temperature difference of 10 K stores 10 K · 4 MJ/m^{3}K = 40 MJ/m^3. If an ice slurry with an ice fraction of 30 vol.% is used, the storage density is increased by 30 % · 330 MJ/m^3 = 99 MJ/m^3. This is an additional 250 %, and therefore means an increase by a factor of 3.5 in the storage density. Assuming the same volume flow for both options this also means an increase of the power by a factor of 3.5.

5.5.1.4 Further information

The discussion of slurries has been very simplified and many of the problems arising when a slurry is used can not be discussed here. Some of them refer to the stability of the slurry regarding capsule tightness, performance of emulsifiers, etc. Ice slurries are different, because they consist of only one component. A general problem using slurries originates from stratification: PCM and heat transfer fluid have different densities and layers with higher PCM contend can be formed by stratification. These layers have a higher viscosity and pumping will become more difficult. In the worst case, this effect can stop the system from working at all. There is also still considerable work on the determination of thermophysical properties and fluid properties, like the viscosity. Last, but not least, also the technical equipment like pumps and heat exchangers is still under investigation. Ice slurries are also different in this respect: they are commercial for many years and the technical equipment is well developed.

In contrast to the loading and unloading of the storage with pumping the slurry, as shown in fig.5.21, systems have also been proposed where the slurry remains in the storage tank at any time and where heat is exchanged via a heat exchanger. Further on, because of the stratification effect it is possible to separate the PCM from the heat transfer fluid and to pump only the heat transfer fluid out of the storage tank. This avoids problems with the PCM fraction in the distribution system. In Japan, large systems are operated with huge storage tanks with ice slurry for cold storage, and often only cold water and no ice slurry is removed from the storage tank for air-conditioning purposes. An example is an installation in the Kyoto station building.

For those looking for more information, Inaba 2000 discusses the different slurry types in more detail, describes how they are technically generated, their flow behavior and main characteristics when pumped through a pipe, the storage of the slurry, and the heat exchange and heat transfer. Egolf et al. 2003a further on discuss advantages, disadvantages, and possible applications of the different types of slurries. Egolf et al. 2006 discuss the problem of stratification on a theoretical basis.

Ice slurries have been investigated and applied for a long time, and numerous publications and review articles specialized on them are available. Egolf et al. 2003b presented a review on physical properties of ice slurries and their industrial applications, including a lot of technical information. Egolf and Kauffeld 2004 also give a short review of the basic research on ice slurries. Furthermore, practical problems of the application of the technology in refrigeration and process techniques are discussed. Egolf 2004 describes different technologies to produce ice and ice slurries. Dincer and Rosen 2002 also describe ice slurries, ice making by ice harvesters and ice on coil technique, and some application examples. The working party on ice slurries of the International Institute of Refrigeration (IIR) has organized a series of workshops and published proceedings like the Proceedings of the 5th Workshop on Ice Slurries of the IIR (Melinder 2002) and the Proceedings of the 6th Workshop on Ice Slurries of the IIR (Kauffeld et al. 2005a). There is also a Handbook on Ice Slurries - Fundamentals and Engineering by Kauffeld et al. 2005b.

Slurries from microencapsulated PCM can be produced with melting temperatures covering a wide range, for cooling as well as for heating applications. More information on microencapsulated PCM slurries can be found in Jahns 2003 and Gschwander et al. 2004. Gschwander and Schossig 2004 describe test results of a microencapsulated PCM slurry in a pipe system including heat exchanger and pump. Heinz and Streicher 2005 and Heinz and Streicher 2006 describe the experimental testing of a storage tank filled with a microencapsulated PCM slurry.

5.5.2 Sensible liquid type

In the limit of vanishing PCM fraction, the heat storage only contains the heat transfer fluid. Because there is no phase change anymore, the heat transfer fluid only stores sensible heat. Therefore, the heat transfer fluid is always a liquid; gases do not store significant amounts of heat. By far the most common example of this type, and probably of heat storage at all, is using water as sensible heat storage medium. Hot water heat storages for temperatures up to about 200 °C are in use. Cold water or chilled water storages to store cold are used in the range of 5 °C to 15 °C. Other heat transfer fluids, which are also used as storage medium, are water-salt or water-alcohol mixtures for temperatures below zero, and oils and salt mixtures for temperatures up to 300 °C.

This type of storage is not a latent heat storage anymore and therefore not part of the topic of this book. It is discussed here because it completes the set of different storage types in a logic way and because it is usually the reference case when different storage types are compared.

5.5.2.1 Construction principle and typical performance

The most common construction principle is for direct discharge, as shown in fig.5.23. In this type, the heat transfer fluid is stored in the storage tank at the desired or a somewhat higher temperature as necessary for the heat demand and when heat is needed, the heat transfer fluid is discharged directly to the demand. The direct discharge is only possible because heat transfer fluid and storage medium are identical.

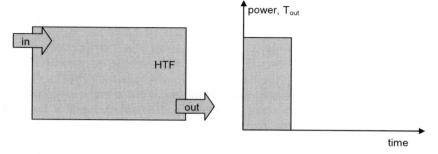

Fig. 5.23. Heat storage using sensible heat only and direct discharge: principle and general performance.

Heat storages using sensible heat only and direct discharge typically have:

- the lowest storage density, compared to other storage types, because only sensible heat storage is used, and
- high and constant power, restricted only by the discharge volume flow.

In the description in fig.5.23 effects of mixing, stratification, and heat losses are not considered, so the temperature at the outlet is equal to the maximum temperature in the storage. Further on, even when heat storage is only by sensible heat in a fluid, it is possible that the heat transfer fluid is a different one; in that case, a heat exchanger has to be used at the inlet and / or outlet.

5.5.2.2 Example

A very common example for a heat storage using sensible heat only and direct discharge is a storage for domestic hot water. To fill a bathtub for example, water with a temperature of at least 40 °C at a flow rate of 12 liters per minute (0.2 liters per second) or more is typically required. This is equivalent to a heating power of about 24 kW. Instead of heating the water just when needed, and therefore to avoid the large heating power, the water is heated long before it is needed at a low heating power and then stored in a storage tank until it is needed. The disadvantage of the low storage density is here balanced by the advantage of a high power that is only restricted by the flow rate.

5.5.2.3 Heat transfer calculation

The heat transfer calculations for a heat storage using sensible heat only and direct discharge is quite simple. The *heat stored* is

$$Q = V^{HTF} \cdot \rho^{HTF} \cdot c_p^{HTF} \cdot (T_{max} - T_{in}) \qquad (5.28)$$

and the *power supplied* is given by

$$P = \frac{d}{dt}Q = \frac{d}{dt}V^{HTF} \cdot \rho^{HTF} \cdot c_p^{HTF} \cdot (T_{max} - T_{in}). \qquad (5.29)$$

For simplification, effects of heat loss, stratification, and mixing have been neglected. It was assumed that the storage is fully loaded with the liquid at T_{max}, and that at discharge the fluid still has the same temperature. This simplification is necessary to be able to compare different storage types at an introductory level, it is however not typical for many applications. Stratification and mixing are often crucial for the performance of such storages. The fact that the density of water decreases with rising temperature (above 4 °C) causes the water with the highest temperature, and thus lowest density, to accumulate at the top of a storage tank. Layers of water with a temperature increasing from the bottom to the top will form this way, just driven by the free convection due to the temperature dependent density. Unless forced convection causes too much mixing, the effect of stratification can be used to store water of different temperatures in the same storage tank and with varying volume fractions. Especially in solar heating systems this can be a significant advantage, because depending on the solar irradiation water at different temperatures is supplied to the storage by the solar collector. With stratification, water supplied at a higher temperature than the average temperature in the storage will be stored at the top of the storage tank and is available for the user. With mixing, water supplied at a higher temperature is mixed with the colder water in the storage. The maximum water temperature in the tank is therefore reduced by mixing. In the worst case, after mixing the temperature is so low that no water at a temperature sufficiently high for the user is available.

5.5.2.4 Further information

The sensible liquid type is not a latent heat storage anymore and therefore not described in any detail here. The interested reader can get more information, especially on hot water heat storage, in the books of Dincer and Rosen 2002, and Streicher and Bales 2005. Dincer and Rosen 2002 for example give a good overview on different tank systems using stratification. An important topic for the design and evaluation of storages are characteristic numbers. Dincer and Rosen 2002 give a detailed description on stratification in sensible heat storages including characteristic numbers for their evaluation. Further on, they discuss 1-dimensional and 2-dimensional models. The numerical modeling of a storage, as well as of a storage in a system like a house, can be performed using the software TRNSYS.

For TRNSYS, several models for sensible heat storages with different construction and restrictions are available.

When taking advantage of the effect of stratification, it is possible to improve a sensible liquid storage, like a hot water heat storage, by the addition of PCM modules in a certain part of the storage. In hot water heat storages with stratification, the lower layers are subject to large temperature changes depending on the loading state of the storage, while the top layers are most of the time close to the demand temperature. Therefore, PCM-modules lead to a more significant improvement of the thermal performance if placed in the top layers; the lower layers already store considerable sensible heat due to the large temperature changes. Mehling et al. 2002 and Mehling et al. 2003 first presented this concept together with experimental and numerical investigations. The numerical model used is based on a packed bed model and can be used as an introductory example for that topic. Cabeza et al. 2004 describe and discuss results of numerical simulations of such a system and experimental results from a real system that compares a tank with PCM module and a tank without. A description of this system contains section 9.3.5.2.

5.6 References

[Arkar and Medved 2002] Arkar C., Medved S.: Enhanced Solar Assisted Building Ventilation System Using Sphere Encapsulated PCM Thermal Heat Storage. Presented at 2nd Workshop IEA ECES Annex 17 "Advanced thermal energy storage techniques – Feasibility studies and demonstration projects", Ljubljana, Slovenia, 3-5 April 2002. www.fskab.com/annex17

[Arkar and Medved 2005] Arkar C., Medved S.: The influence of the thermal properties of a PCM on the accuracy of a numerical model of a packed bed latent heat storage with spheres. Presented at 8th Workshop IEA ECES Annex 17 "Advanced thermal energy storage techniques – Feasibility studies and demonstration projects", Kizkalesi, Turkey, 18-20 April 2005. www.fskab.com/annex17

[Bédécarrats et al. 1996] Bédécarrats J.P., Strub F., Falcon B., Dumas J.P.: Phase-change thermal energy storage using spherical capsules: performance of a test plant. Int. J. Refrig, Vol.19 No.3, 187-196 (1996)

[Bédécarrats and Dumas 1997] Bédécarrats J.P., Dumas J.P.: Etude de la cristallisation de nodules contenant un materiau à changement de phase en vue du stockage par chaleur latente. Int. J. Heat Mass Transfer, Vol.40, No.1, 149-157 (1997)

[Cabeza et al. 2004] Cabeza L.F., Nogués M., Roca J., Ibañez M.: PCM research at the University of Lleida (Spain). Presented at 6th Workshop IEA ECES Annex 17 "Advanced thermal energy storage techniques – Feasibility studies and demonstration projects", Arvika, Sweden, 8-9 June 2004. www.fskab.com/annex17

[Çengel 1998] Çengel Y.: Heat transfer – A practical approach. Mc Graw Hill (1998)

[Dincer and Rosen 2002] Dincer I., Rosen M.A.: Thermal energy storage. Systems and applications. John Wiley & Sons, Chichester (2002)

[Dolado et al. 2006] Dolado P., Lázaro A., Zalba B., Marín J.M.: Numerical simulation of the thermal behaviour of an energy storage unit with phase change materials for air conditioning applications between 17 °C and 40 °C. Proc. of ECOSTOCK, 10th International Conference on Thermal Energy Storage, Stockton, USA, 2006

[Egolf and Manz 1995] Egolf P., Manz H.: Latentwärmespeicher für Wassersysteme. HLH, Bd. 46, 499 – 503 (1995)

[Egolf et al. 2003a] Egolf P.W., Sari O., Kitanovski A.: Multifunctional Thermal Fluids for Refrigeration and Air Conditioning. Proc. of 10th European conference on technological Innovations in air conditioning and refrigeration industry, 2003, http://www.centrogalileo.it/nuovaPA/Articoli%20tecnici/INGLESE%20CONVEGNO/INDICE.htm

[Egolf et al. 2003b] Egolf W., Sari O., Vuarnoz D., Caesar D.A., Sletta J.: A review from physical properties of ice slurry to industrial applications. Proc. of Phase change material & slurry engineering conference & business forum, Yverdon-les-Bains, Switzerland, 2003

[Egolf 2004] Egolf P.W: Ice slurry: a promising technology. Technical note on refrigerating technologies, www.iifiir.org, 2004

[Egolf and Kauffeld 2004] Egolf P.W., Kauffeld M.: From physical properties of ice slurries to industrial ice slurry applications. Int. J. of Refrig., Vol. 28, Issue 1, 4 - 12 (2005)

[Egolf et al. 2006] Egolf P.W., Saraswat M., Bajpai H., Sari O.: Toward a flow phase diagram of ice slurry. Proc. of 7th Conference on Phase Change Materials and Slurries for Refrigeration and Air Conditioning, Dinan, France, 13 - 15 September 2006

[Farid and Yacoub 1989] Farid M.M., Yacoub K.: Performance of direct contact latent heat storage unit. Solar energy **43**, 237-252 (1989)

[Farid et al. 2004] Farid M.M., Khudhair A.M., Razack S.A.K., Al-Hallaj S.: A review on phase change energy storage: materials and applications. Energy Conversion and Management **45**, 1597–1615 (2004)

[Fouda et al. 1984] Fouda A.E., Despault G.J., Taylor J.B., Capes C.E.: Solar storage system using salt hydrate latent heat and direct contact heat exchange – II, characteristics of pilot operating with sodium sulphate solution. Solar energy **32**, 57-65 (1984)

[Gschwander et al. 2004] Gschwander S., Schossig P., Henning H.-M.: Mikroverkapselte Phasenwechselmaterialien in Fluiden zur Erhöhung der Wärmekapazität. Proc. of 14. Symposium Thermische Solarenergie, OTTI Technologie-Kolleg, Staffelstein 2004, pp. 512-516.

[Gschwander and Schossig 2004] Gschwander S., Schossig P.: First Results testing a microencapsulated PCM−Slurry. Presented at 6th Workshop IEA ECES Annex 17 "Advanced thermal energy storage techniques – Feasibility studies and demonstration projects", Arvika, Sweden, 8-9 June 2004. www.fskab.com/annex17

[He et al. 2004] He B., Rydstrand M., Pettersson J., Martin V., M. Westermark: The first-step experiments on dynamic process phase change material (PCM) storage system for comfort cooling applications. Presented at 7th Workshop IEA ECES Annex 17 "Advanced thermal energy storage techniques – Feasibility studies and demonstration projects", Beijing, China, 2004. www.fskab.com/annex17

[Heinz and Streicher 2005] Heinz A., Streicher W.: Experimental testing of a storage tank filled with microencapsulated PCM slurries. Proc. of 2nd Phase Change Material and Slurry Scientific Conference & Business Forum, Yverdon-les-Bains, Switzerland, June 15-17, 2005

[Heinz and Streicher 2006] Heinz A., Streicher W.: Application of phase change materials and PCM-slurries for thermal energy storage. Proc. of ECOSTOCK, 10th International Conference on Thermal Energy Storage, Stockton, USA, 2006

[Heinz 2007] Heinz A.: Application of Thermal Energy Storage with Phase Change Materials in Heating Systems. PhD thesis at the Institut für Wärmetechnik, TU Graz, 2007

[Inaba 2000] Inaba H.: New challenge in advanced thermal energy transportation using functionally thermal fluids. Int. J. Therm. Sci.**39**, 991-1003 (2000)

[Ismail and Stuginsky 1999] Ismail K.A.R., Stuginsky R.: A parametric study on possible fixed bed models for pcm and sensible heat storage. Applied Thermal Engineering **19** (7), 757-788 (1999)

[Jahns 2003] Jahns E.: Microencapsulated phase change slurries. Proc. of Phase Change Material and Slurry Scientific Conference & Buisiness Forum; Yverdon-les-Bains, 2003

[Kauffeld et al. 2005a] Kauffeld M., Egolf P.W., Sari O. (eds): Proceedings of the 6th Workshop on Ice Slurries of the IIR; IIF-IIR, France, 2005

[Kauffeld et al. 2005b] Kauffeld M., Kawaji M., Egolf P.W., Melinder A., Davies T.W.: Handbook on Ice Slurries - Fundamentals and Engineering. IIF-IIR, France, 2005, ISBN 2-913149-42-1

[Kenisarin and Mahkamov 2007] Kenisarin M., Mahkamov K.: Solar energy storage using phase change materials. Renewable and Sustainable Energy Reviews, Vol.11 (9), 1913-1965 (2007)

[Lamberg and Siren 2003a] Lamberg P., Siren K.: Analytical model for melting in a semi-infinite PCM storage with an internal fin. Heat and Mass Transfer **39**, No. 2, 167-176 (2003)

[Lamberg and Siren 2003b] Lamberg P., Siren K.: Approximate analytical model for solidification in a finite PCM storage with internal fins. Applied Mathematical Modelling **27**, No. 7, 491-513 (2003)

[Lamberg 2004] Lamberg P.: Approximate analytical model for two-phase solidification problem in a finned phase-change material storage. Applied Energy **77**, No. 2, 131-152 (2004)

[Lindner 1984] Lindner F.: Latentwärmespeicher Teil 1: Physikalisch-technische Grundlagen. Brennst.-Wärme-Kraft **36**, Nr. 7-8 (1984)

[Mehling 2001] Mehling H.: Latentwärmespeicher – Neue Materialien und Materialkonzepte. Proc. of FVS Workshop 'Wärmespeicherung', Cologne, Germany, May 2001. http://www.fv-sonnenenergie.de/Publikationen/index.php?id=5&list=23

[Mehling et al. 2002] Mehling H., Cabeza L., Hiebler S., Hippeli S.: Improvement of stratified hot water heat stores using a PCM-module. Proc. of EuroSun Bologna, June 2002

[Mehling et al. 2003] Mehling H., Cabeza L.F., Hippeli S., Hiebler S.: PCM-module to improve hot water heat stores with stratification. Renewable Energy **28**, 699–711 (2003)

[Melinder 2002] Melinder A. (ed): Proc. of the 5th Workshop on Ice Slurries of the IIR; IIF-IIR, France, 2002

[Naumann et al. 1989] Naumann R., Eildermann G., Günther A.: Latentwärmespeicher – Entwicklung und chemische Probleme. Zeitschrift für Chemie, Heft 10, October 1989, ISSN 0044-2402

[Pluta 2003] Pluta Z.: Numerical analysis of pebble/PCM bed storage systems for solar air heating installations. Proc. of Futurestock, 9th International Conference on Thermal Energy Storage, Warsaw, Poland, 2003

[Streicher and Bales 2005] Streicher W., Bales C.: Combistores. In Hadorn J.-C. (ed.) Thermal energy storage for solar and low energy buildings, pp. 29-39. Servei de Publicacions (UDL), Lleida (2005) ISBN 84-8409-877-X

[Wang et al. 2003] Wang H., Chen C., Jian R., Li X.: Discussion on Comparison of Numerical Calculating Methods on Phase-change Heat Transfer Problems. Proc. of Futurestock, 9th International Conference on Thermal Energy Storage, Warsaw, Poland, 2003

[Xiao et al. 2006] Xiao R., Wu S., Tang L., Huang C., Feng Z.: Experimental investgaton of the pressure drop of clathrate hydrate slurry (CHS) flow of Tetra Butyl Ammonium Bromide (TBAB) in straight pipe. Proc. of ECOSTOCK, 10th International Conference on Thermal Energy Storage, Stockton, USA, 2006

[Zalba et al. 2003] Zalba B., Marin J.M., Cabeza L.F., Mehling H.: Review on thermal energy storage with phase change materials, heat transfer analysis and applications. Applied thermal Engineering **23**, 251 – 283 (2003)

6 Integration of active storages into systems

In chapter 5, the different storage types including their construction principle, an example, and a simple heat transfer calculation were discussed. Active storages, that means storages where the heat transfer fluid is actively moved, allow a full control during charging and discharging. The option to control the flow of the heat transfer fluid and therefore the heat transfer in and out of the storage allows optimizing the performance of a storage within a system, or better the overall system performance. On the other hand a well-designed storage, which is badly integrated into a system, will finally be ineffective. How are active storages integrated into a system? Are there general concepts? This is the topic of this chapter: an introduction into the basics of integrating active storages into a system.

6.1 Integration goal

Before discussing the integration concepts, it is better to remember again what the goal is. The main goal to integrate a heat or cold storage into a system is of course to supply heat or cold. However, there are different supply and demand situations that have a great influence on the integration concept. The easiest case with no overlap in time between loading from the supply and unloading to the demand side shows fig.6.1. The goal is to match different times of supply and demand, and in many cases different supply and demand powers. If there is no overlap in time, even transport from one place to another is possible.

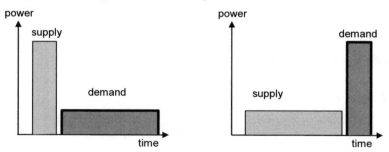

Fig. 6.1. Two cases with no overlap in time between supply and demand.

If there is a partial or total overlap in time, as shown in fig.6.2, it is additionally possible to smooth out fluctuations of the supply or of the demand. Typical goals of a storage integration are temperature regulation and power matching in a grid where supply and demand have a fixed connection.

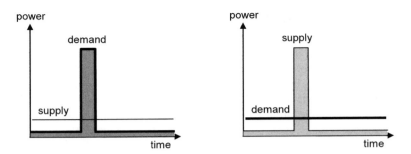

Fig. 6.2. Two cases with overlap in time between supply and demand.

The basic goals of the storage are therefore to match supply and demand regarding

- time,
- power,
- location.

Potential advantages on the overall system performance are

- better economics, that is reduced investment and running cost
- better efficiency, that means more efficient use of energy
- less pollution of the environment, for example by CO_2
- better system performance and reliability

The focus here is on showing that a set of different integration concepts exists. They all have different advantages and disadvantages with respect to their contribution to the performance of a system.

6.2 Integration concepts

6.2.1 General concepts

The basic goals of the storage are to match supply and demand regarding the amount of heat or cold and the heating or cooling power at the right time. While the amount of heat or cold is determined by the size of the storage, and the heating or cooling power mainly by the design of the storage, the integration concept has a large influence with respect to time. Fig.6.3 shows the basic integration concepts of storages.

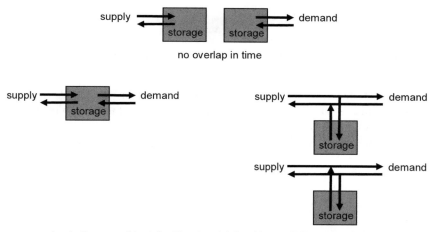

allways overlap in time; left upstream operation, right downstream operation

Fig. 6.3. Basic concepts to integrate storages into systems, and their influence on the performance with respect to time. The arrows indicate the flow of a heat transfer fluid.

The top of fig.6.3 shows the concept for a system with no overlap between supply and demand, which means where the function is storage mainly. This includes the case of a decentralized system where supply and demand are at different locations. First, the storage is loaded from the supply and disconnected; later it is connected to the demand an unloaded. The middle shows the three integration concepts where it is possible, but not necessary, to load and unload the storage at the same time. At the left, this is done with two independent circuits; at the right with the same one. These concepts can do both, storage and buffering. The bottom shows the two integration concepts with supply, demand, and storage in series. For these concepts, there is always an overlap in time. In upstream operation, the supply is directly connected to the demand and therefore any change in the supply will act directly on the demand. The function of the storage is to buffer the feedback of changes in the return from the demand side to the supply side. In downstream operation, any change in the supply will be buffered by the storage before the heat transfer fluid is supplied to the demand side.

6.2.2 Special examples

When planning the integration of a storage into a system there is a series of technical questions regarding the heat transfer to and from the heat storage:

- is the heat transfer through a solid by heat conduction, or by a fluid?
- if it is a fluid, is it moved actively for the transfer of heat to reach higher power or to control the power?
- which heat transfer medium is used? Is it the same on the supply and demand side?
- if it is actively moved, is it also circulating within the storage volume to allow a larger heat transfer area than if only the surface area of the storage volume is used?

In this respect, fig.6.3 is very simplified. The arrows only indicate the flow of a heat transfer fluid in and out on the demand and supply side. However, how the heat is transferred to or from the heat transfer fluid to the storage material is not indicated. The basic concepts for this have been discussed in chapter 5, however with a simplification: the concepts can be used on both sides and in different combinations. Fig.6.4 shows a few examples for the basic concept from fig.6.3 for possible overlap in time, but without the possibility to bypass the storage. While on the supply side the heat transfer is always active and with a heat exchanger, the method for heat transfer on the demand side is now varied. The first variation at the top is with heat transfer at the surface of the storage; this could be to air, or to another solid. Then follow three variations with heat transfer on internal surfaces: by a heat exchanger, with PCM modules, and by direct contact to the storage medium. The last variation at the bottom is by exchanging the storage medium, a slurry.

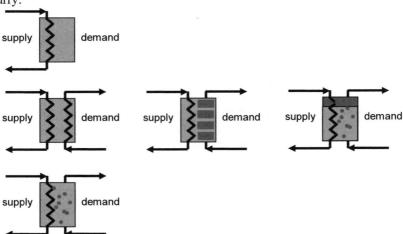

Fig. 6.4. Variations of the method of heat transfer on the demand side.

Of course, the same variations can also be applied to the supply side. This results in a large variety of concepts. When making different combinations, like having a slurry in a tank that is discharged directly, but heated via a heat exchanger, the notation for storage types developed in section 5.2 is not strictly applicable. The reason is that the notation is based on a single approach for heat transfer. In cases where two approaches have been combined they have to be mentioned both to avoid ambiguity.

In a whole system, the situation can be even more complicated. Up to now, the heat transfer fluid in a storage was identical at least with the heat transfer fluid from the supply or the demand side connection. Sometimes however there is a technical advantage to use a different heat transfer fluid. An example is a direct contact LHS with a salt hydrate as storage medium and oil as internal heat transfer medium, which is loaded with hot water from a solar collector and unloaded to a water based heating system. This case, where the water circuits on the demand and supply side are separated from the oil circuit that loads and unloads the LHS is schematically shown on the left in fig.6.5.

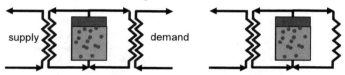

Fig. 6.5. Integration concepts with a secondary heat transfer fluid at the supply and demand side.

The right side shows a system where the heat on the demand side is transferred from the heat exchanger to air. By using secondary heat transfer fluids it is possible to separate many restrictions on the heat transfer fluid given by the demand and supply side from the storage design. However, the additional heat exchangers use part of the available temperature difference between supply and demand.

6.3 Cascade storages

The main advantage of a latent heat storage compared to a sensible heat storage is the high storage density at small temperature differences. This does not automatically mean that latent heat storage cannot be useful also at large temperature differences; it is still possible to divide a larger temperature difference into smaller ones, each covered with a separate PCM with a different phase change temperature. However, what exactly is the advantage then?

The example in fig.6.6 shows a sensible heat storage compared to a latent heat storage with a single PCM. For the small temperature difference, there is a factor of 3 between the heat stored in the latent and the sensible heat storage. With increasing temperature difference this advantage becomes smaller and smaller. For the large temperature difference indicated, there is only a ratio of 6:4 = 1.5 between latent and sensible heat storage.

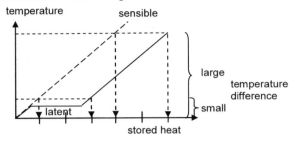

Fig. 6.6. Comparison of a sensible heat storage and a latent heat storage with a single PCM.

The use of several PCM with different melting temperatures seems to solve this problem. Such a combination with different melting temperatures is called a *cascade storage*. Fig.6.7 shows a cascade storage built using three different PCM with different melting temperatures.

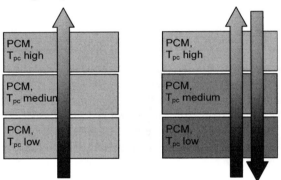

Fig. 6.7. Cascade storage with three heat exchanger type LHS in series. Left: only one heat exchanger to load and unload. Right: with separate heat exchangers and counter flow.

However, the situation is more complex because the boundary conditions have a large influence on the storage performance. To explain this, two different cases are now discussed. The first case is when the storages are loaded from a single inlet temperature to a single maximum temperature, the same way as discussed throughout chapter 5. Fig.6.8 shows the heat stored for the sensible and for the cascade heat storage. The three latent heat storages with different melting temperatures are all heated from the same inlet temperature to the same maximum temperature. To have the same volume as in the cascade heat storage, the stored heat in the latent heat storage is added three times. Then, the ratio of the stored heat is 18:12 = 1.5; this is the same result as before and rather disappointing.

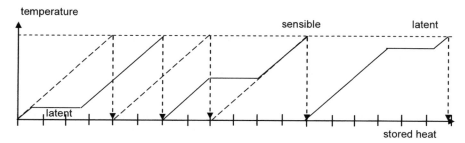

Fig. 6.8. Comparison of a sensible heat storage and three latent heat storages with three different PCM for the same ΔT for each storage.

The result becomes different if the storages do not use the same temperature interval. Fig.6.9 shows this case. The LHS with the low melting temperature is heated from T_1 to T_2, the LHS with the medium melting temperature is heated from T_2 to T_3, and the LHS with the high melting temperature is heated from T_3 to the maximum temperature.

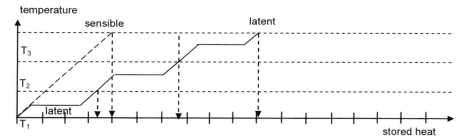

Fig. 6.9. Comparison of a sensible heat storage and three latent heat storages with three different PCM using different temperature intervals.

In this scenario, the LHS loose only little storage density because the temperature ranges for latent heat storage are still fully used. For the latent heat storage, the consequence of using the small temperature intervals is large. If the sensible heat storage is installed the same way as the cascade storage, it consists of three storages all heated only by a third of the total temperature range. Therefore, the difference in the storage capacities in this scenario is 10:4 = 2.5; this is almost the same as for the case of a small temperature interval and a single melting temperature. This is an impressive result, and raises the question when this scenario is applicable. There are two cases:

- The first case is when the times for loading and unloading the storage are limited. When loading a storage with a single PCM, the heat transfer fluid rapidly transfers heat to the PCM when it enters the storage. This reduces the temperature of the heat transfer fluid, therefore also the temperature gradient to the PCM, and therefore the heat transfer to the PCM at the end of the storage.

Consequently, where the heat transfer fluid enters the storage the PCM is melted fast, at the end of the storage it is melted slower. For the discharging process, this is shown in fig.5.13. When the time for loading or unloading is too short, the PCM at the end of the storage therefore might not be used for latent heat storage. What is the consequence of this effect on a cascade storage? The above discussion was based on the assumption that all storages use the full temperature range. Assuming now a limited time for loading and unloading, this might not be the case. With a cascade storage this problem can be solved. Using a PCM with a lower temperature at the end of the heat exchanger the temperature gradient in the loading process can be high enough to load the full storage. If discharge is done the opposite direction, as shown on the right in fig.6.7, this also works for discharge.

- The second case is derived from the first case: the storage density of the cascade storage is more evenly distributed over the temperature range, similar to a sensible heat storage with a higher heat capacity. This means that heat at different temperature levels can now be stored with a high storage density.

Michels and Pitz-Paal 2007 have investigated the approach of a cascade LHS for parabolic trough solar power plants experimentally and numerically. The goal was to minimize the amount of storage material necessary to achieve the necessary storage capacity. If the advantage of using a combination of different latent heat storages is large enough to accept the significant disadvantage of a more complex storage and system design depends on the actual application.

6.4 Simulation and optimization of systems

To model a latent heat or cold storage in a whole system, it is necessary to adapt the numerical description of the storage to a software that is capable of simulating a system. This is often a complicated task and solutions for many problems have just been developed in recent years. For example for applications to heat and cool buildings it has been a goal for a long time to develop a numerical code for a LHS that can be implemented into a numerical software for building simulations, like TRNSYS. Puschnig et al. 2005 and Bony et al 2005 present several simulation models with different levels of simplification for storages filled with a PCM slurry and / or PCM modules. For the software DYMOLA MODELICA, Brun et al. 2006 have developed a description for a slurry system.

If a simulation tool is available for a whole system, it is possible to discuss different strategies for the optimization of the system with regard to energy and exergy, energy savings, demand charges, reduction of equipment, and storage design. Information on these topics is available in the publications by Dincer and Rosen 2002 and ASHRAE 1991.

6.5 References

[ASHRAE 1991] ASHRAE Handbook, HVAC Applications, SI Edition. ASHRAE Atlanta (1991), ISBN 0-910110-79-4

[Bony et al. 2005] Bony J., Ibáñez M., Puschnig P., Citherlet S., Cabeza L., Heinz A.: Three different approaches to simulate PCM bulk elements in a solar storage tank. Proc. of 2nd Conference on Phase Change Material and Slurry, Yverdon-les-Bains, 2005

[Brun et al. 2006] Brun F., Gendre F., Siegenthaler E., Sari O., Egolf P.W.: Phase change slurry system design with DYMOLA MODELICA. Proc. of 7th Conference on Phase Change Materials and Slurries for Refrigeration and Air Conditioning, Dinan, France, 13 - 15 September 2006

[Dincer and Rosen 2002] Dincer I., Rosen M.A.: Thermal energy storage. Systems and applications. John Wiley & Sons, Chichester (2002)

[Michels and Pitz-Paal 2007] Michels H., Pitz-Paal R.: Cascaded latent heat storage for parabolic trough solar power plants. Solar Energy, Vol. 81, Nr. 6, June, 829-837 (2007), doi:10.1016/j.solener.2006.09.008

[Puschnig et al. 2005] Puschnig P., Heinz A., Streicher W.: TRNSYS simulation model for an energy storage for PCM slurries and / or PCM modules. Proc. of 2nd Conference on Phase Change Material and Slurry, Yverdon-les-Bains, 2005

7 Applications in transport and storage containers

When preparing a barbeque on a hot summer day, it is usually necessary to keep beverages and food cold. An easy way to do this is to put an ice pack into the transport and storage containers that are used for the food and beverages. This is very convenient, no connection to the electricity grid is necessary, and it makes no noise. In the past decade, the application of PCM in transport containers became one of the first fully commercial PCM applications. Examples are the transport of food, medicine, and pharmaceuticals. They all have in common that it is important to keep the temperature of the product above or below a certain temperature, or within a narrow temperature range. Further on, in many cases like in the cargo bay of an airplane, a connection to the electricity grid to use a cooler or a heater is not available. In these cases, PCM technology offers a unique solution.

In this chapter, first the basics and then examples for different applications in transport and storage containers are discussed. Applications in transport containers are applications where the focus is on the temperature control and not the supply of larger amounts of heat or cold. Therefore, usually no internal heat transfer fluid is used within the PCM; the heat exchanged at the surface is sufficient. In this first chapter on applications, it is therefore possible to apply the very basic calculations of heat transfer as laid out in chapter 4 and section 5.3.

7.1 Basics

In the following section, the theoretical basis for temperature control is derived, starting with discussing the cool down of the object, which temperature has to be controlled, and ending with the object and PCM module in an insulated environment.

7.1.1 Ideal cooling of an object in ambient air

Fig.7.1 shows the case where the object, which temperature has to be controlled, looses heat to the ambient air and cools down.

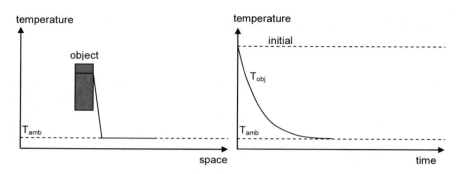

Fig. 7.1. Cooling of an object, which stores only sensible heat, in ambient air.

At this point, it is assumed that the object, which temperature has to be controlled, stores only sensible heat. Further on, to simplify the heat transfer calculations, it is also assumed that

- the object is allways isothermal,
- the heat transfer coefficient object-air is constant, and
- the temperature of the ambient air is constant.

Under these assumptions the cool down of the object in ambient air can be described. The heat lost by the object is equal to the heat transferred to the ambient

$$\frac{d}{dt}Q = m \cdot c_p \cdot \frac{d}{dt}T_{obj} = -A_{obj} \cdot \alpha \cdot (T_{obj} - T_{amb}). \tag{7.1}$$

This gives the simple differential equation for the temperature of the object

$$\frac{d}{dt}T_{obj} = -\frac{A_{obj} \cdot \alpha}{m \cdot c_p} \cdot (T_{obj} - T_{amb}). \tag{7.2}$$

Because T_{amb} is constant, it is possible to add T_{amb} to the derivative. Then, the differential equation is

$$\frac{d}{dt}(T_{obj} - T_{amb}) = -\frac{A_{obj} \cdot \alpha}{m \cdot c_p} \cdot (T_{obj} - T_{amb}). \tag{7.3}$$

It is of the general form $dx/dt = -a \cdot x$, and has the solution

$$T_{obj}(t) - T_{amb} = (T_{obj}(t=0) - T_{amb}) \cdot \exp\left(-\frac{A_{obj} \cdot \alpha}{m \cdot c_p} \cdot t\right). \tag{7.4}$$

The cool down of the object thus follows an exponential curve with the heat transfer area, the heat transfer coefficient, the mass, and the heat capacity being the dominating parameters.

If the object undergoes a phase change, the situation is different. The latent heat is released at the phase change temperature and consequently, the heat loss is compensated for a time as long as the PCM undergoes the phase transition. After the phase change is completed, the temperature follows the same path as without phase change. As fig.7.2 shows, this simply results in a shift in the temperature-time curve, the same way as in the T-History method (fig.3.17).

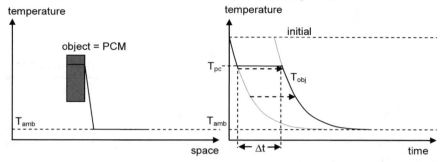

Fig. 7.2. Cooling of an object with sensible and latent heat in ambient air.

Knowing the heat stored in the phase change $m \cdot \Delta_{pc}h$, the heat transfer coefficient α, and the surface area A, the temperature is controlled until the stored heat is lost to the ambient. This means

$$\Delta_{pc}H = m \cdot \Delta_{pc}h = \dot{Q} \cdot \Delta t = A_{obj} \cdot \alpha \cdot (T_{pc} - T_{amb}) \cdot \Delta t \qquad (7.5)$$

and therefore

$$\Delta t = \frac{m \cdot \Delta_{pc}h}{A_{obj} \cdot \alpha \cdot (T_{pc} - T_{amb})} \qquad (7.6)$$

7.1.2 Ideal cooling of an insulated object in ambient air

To slow down the cooling of the object, and thereby stabilize its temperature for a longer time, the heat loss has to be reduced. This can be done by applying an insulating layer, as shown in fig.7.3.

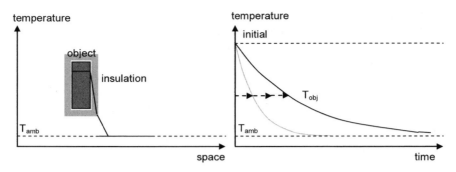

Fig. 7.3. Cooling of an insulated object, which stores only sensible heat, in ambient air.

To keep the calculations simple it is necessary to make the following additional simplifications for the insulation:

- the insulation does not store any heat, and
- the insulation is infinitely thin, so the area of heat transfer remains unchanged.

The insulation thus only replaces the thermal resistance of the air layer in eq.7.1

$$A_{obj} \cdot \alpha \longrightarrow A_{obj} \cdot k_{eff} = A_{obj} \cdot \frac{1}{1/\alpha + 1/k_{ins}} \tag{7.7}$$

with $k_{ins} = \lambda / d$ and d as the thickness of the insulation and λ the thermal conductivity of the insulation material. Using this in eq.7.4 gives

$$T_{obj}(t) - T_{amb} = \left(T_{obj}(t=0) - T_{amb}\right) \cdot \exp\left(-\frac{A_{obj} \cdot k_{eff}}{m \cdot c_p} \cdot t\right)$$

$$= \left(T_{obj}(t=0) - T_{amb}\right) \cdot \exp\left(-\frac{A_{obj} \cdot \frac{1}{1/\alpha + 1/k_{ins}}}{m \cdot c_p} \cdot t\right) \tag{7.8}$$

Because everything else is kept constant, the solution scales with the product $k_{eff} \cdot t$. A typical value for α in air with free convection is about $4 - 8$ W/m²K. For the insulation, a layer of styrofoam with $\lambda = 0.040$ W/mK and thickness of $d = 0.02$ m results in $k_{ins} = 2$ W/m²K. The effective heat transfer coefficient is thus reduced to about 1/3 of its original value and consequently the cool down takes three times as long. This is shown in fig.7.3 on the right.

7.1.3 Ideal cooling of an insulated object with PCM in ambient air

The next step of improving the system is to combine the effect of insulation with the effect of latent heat, as shown in fig.7.4. This results in a slow down of the cooling and at the same time prolongs the stabilization of the temperature of the object.

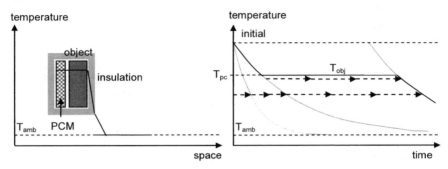

Fig. 7.4. Ideal cooling of an insulated object with PCM. The combination of PCM and insulation keeps the object temperature constant over a long time.

To keep the calculation simple it is necessary to make two additional simplifications:

- the PCM module is also isothermal and at the same temperature as the object, and
- the assumption that the temperature has to be stabilized in such a small temperature range, that the sensible heat stored in the object can be neglected compared to the latent heat stored in the PCM module.

The result is then calculated by combining the two solutions derived before. In eq.7.5, which describes the cool down of the PCM module in ambient air, the heat transfer coefficient is replaced by the one that includes the insulation. Therefore

$$\Delta_{pc} H = m \cdot \Delta_{pc} h = \dot{Q} \cdot t = A_{obj} \cdot k_{eff} \cdot (T_{pc} - T_{amb}) \cdot \Delta t . \qquad (7.9)$$

Then

$$\Delta t = \frac{m \cdot \Delta_{pc} h}{A_{obj} \cdot k_{eff} \cdot (T_{pc} - T_{amb})} . \qquad (7.10)$$

Eq.7.10 is already the result; it describes for how long the temperature of an insulated object with PCM is stabilized. Fig.7.4 shows the solution graphically. The PCM releases additional heat at the phase change temperature, thereby stabilizes the temperature, and slows down the cooling. The insulation prolongs the stabilizing effect of the PCM.

7.1.4 Real cooling of an insulated object with PCM in ambient air

Of course, many of the above assumptions do not strictly hold in reality. The most important deviation is probably that in reality it is very difficult to achieve that PCM, object, and the whole interior of the insulation are at the same temperature. When taking into account the heat transfer coefficients within the insulation, the situation is described on the left of fig.7.5. The PCM, the object, and the interior of the insulation, all have different temperatures.

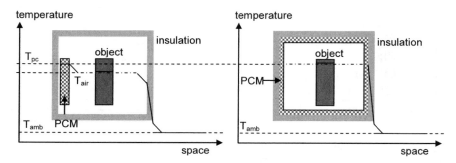

Fig. 7.5. Real cooling of an insulated object. Left: with a PCM module. Right: surrounded with a PCM layer.

To stabilize the temperature of the object as close to the phase change temperature as possible, the PCM can be placed around the internal surface of the insulation, as shown on the right in fig.7.5. During phase change of the PCM, the PCM creates an isothermal enclosure at the phase change temperature. The object surrounded by the PCM is therefore in an isothermal environment at the phase change temperature. The isothermal conditions between PCM and object stop the loss of heat from the object and keep its temperature constant at the phase change temperature. This holds until the phase change is completed.

When the object is a liquid, there is a very common example. Fig.7.6 shows the cooling of a drink with an ice cube.

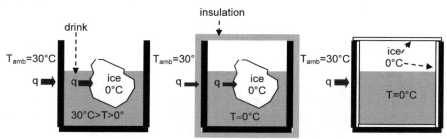

Fig. 7.6. Different methods of keeping a drink cold using ice.

The left side shows the case where only ice is put into the drink. Due to the comparatively large heat flow and the different heat transfer coefficients at the boundaries ice-drink, drink-glass, and glass-ambient, the temperature of the drink will be between the melting temperature of the ice and the temperature of the ambient. When an insulation is applied around the glass, the heat flow is reduced. This in turn reduces the temperature gradients and the temperature of the drink will be closer to 0 °C. When ice plates are used as a kind of "secondary insulation" or "thermal shield", the liquid will cool down to 0 °C and then remain there until the ice is melted.

7.2 Examples

The applications in transport containers cover many diverse fields like the transport or storage of fresh or cooked food, cold or hot beverages, blood derivatives, pharmaceutical products, biomedical products, electronic circuits, and many others. Cabeza et al. 2002 and Métivaud et al. 2005 presented a review on applications in transport containers. Here only a few typical examples are given.

7.2.1 Multi purpose transport boxes and containers

The range of applications for multi purpose transport boxes with controlled temperature is of course much larger than just for food, beverages, and medical applications. Therefore, several companies have developed transport boxes, which are not tailored to any special application. At the left and centre of fig.7.7, an insulated transport box with PCM modules developed by the company PCM Energy P. Ltd is shown. It is available in several sizes, according to PCM Energy P. Ltd. As the pictures show, the PCM is located at the outside of the containment space, as discussed in section 7.1.4.

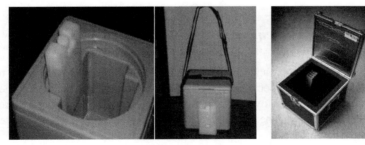

Fig. 7.7. Transport boxes with PCM for different purposes (pictures: left and center PCM Energy P. Ltd, right va-Q-tec AG).

The right side of fig.7.7 shows a box with a highly efficient vacuum super insulation, developed and commercialized by the company va-Q-tec AG. The special insulation has a thermal conductivity of about 4 - 5·10^{-3} W/mK, which is a factor 7 to 8 better than conventional insulation materials. This reduces the necessary cooling power to the order of 10 W, but without significantly reducing the storage space. Tests have proved that using PCM for temperature stabilization, a temperature of -20 °C can be sustained inside the box for 4 days (96 hours) at ambient temperatures of +30 °C. For the transport of pharmaceutical products, it is often required to keep the temperature within +2 °C to +8 °C. In a test, +2 to +8 °C was sustained for 3.5 days (84 hours) at ambient temperatures of above 20 °C.

The company va-q-tec AG also developed a multi purpose transport container for the transport of temperature sensible goods by van, truck, ship, or plane. The container, shown in fig.7.8, combines a highly effective vacuum super insulation with PCM to stabilize the temperature and needs no internal or external electricity source. Its internal dimensions are 1250 mm x 850 mm x 1320 mm and it can fit a content of up to 500 kg.

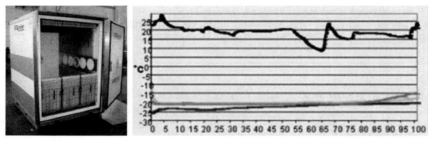

Fig. 7.8. Transport container, which can sustain temperatures of -18 °C at ambient temperatures of around 20 °C for 4 days (pictures: va-q-tec AG).

Fig.7.8 shows measurements from a transport where the temperature in the container was even kept as low as -18 °C for 4 days, while the ambient temperature was around 20 °C.

7.2.2 Thermal management system

As mentioned before, one approach to attain a constant temperature close to the PCM temperature within a container is to place the PCM at the walls. Another option is to have the PCM away from the walls and use forced convection to reduce temperature differences. The company ACME has developed such an active system for larger containers.

In India, there is a rapidly growing need for an energy-efficient and low-cost solution to control the quality and freshness of temperature-sensitive goods. As a solution, ACME has developed containers that can be transported with normal vehicles, thereby eliminating the requirement for expensive refrigerated trucks. Each container is equipped with rechargeable PCM profiles, as shown in fig.7.9. The system has a ventilator on the top to enforce convection.

Fig. 7.9. Thermal management system developed by ACME (picture: ACME)

The PCM profiles are periodically charged through a main unit. Once charged, the desired temperature inside the container can be maintained for up to 48 h. That means the system also allows cooling without running the engine of the truck.

7.2.3 Containers for food and beverages

In catering, meals are prepared at one location and have to be transported to another place where they are eaten. These meals can be hot meals like many main dishes or cold meals like salads, frozen deserts, or ice cream. One example of such an application, which has already been commercialized, is the pizza-heaters shown on the left in fig.7.10. The pizza heater, developed by the company Rubitherm Technologies GmbH, consists of a plate impregnated with PCM. The plate with the pizza on top is put into the transport box for pizza delivery. The PCM pizza heater assures that the pizza is kept above 65 °C three times as long as without, and thereby assures that the pizza can be delivered hot. A second product to keep food warm is shown at the centre of fig.7.10. It is plates fabricated from aluminum profiles and filled with the granulate Rubitherm GR.

Fig. 7.10. Pizza heater, plate for heating and cooling food (pictures: Rubitherm Technologies GmbH) and trolley for food (picture: va-Q-tec)

Another example has been developed by the company va-Q-tec. It is a trolley used to distribute food in hospitals, schools, etc. (fig.7.10, right). According to the respective product data sheet available from the website, the trolley allows storage below -10 °C for almost 35 hours at outside temperatures of 25 °C.

What works for food can also be applied for beverages. The "isothermal bottle", shown in fig.7.11, is a double wall bottle where the PCM fills the space between these walls. The bottle, developed by the company Sofrigam / France, can hold about 0.5 liters of a beverage. The bottle has to be precooled, for example in the refrigerator, to solidify the PCM. Then the bottle will keep the beverage cold at about 13 °C for 3 hours at an ambient temperature of 25 °C (Métivaud et al. 2005).

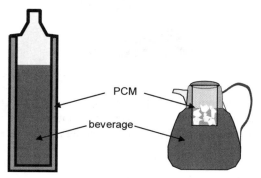

Fig. 7.11. Isothermal bottle developed by the company Sofrigam (left), and vessel to prepare cold drinks used in the past century (right).

The right side of fig.7.11 shows a vessel used in the past century to prepare cold drinks. It consists of a standard vessel to keep the beverage, and an additional vessel containing ice inserted from the top. This way, the drink was cooled with ice, but without the melt water diluting the drink.

In the same way as beverages are cooled, it is of course also possible to heat them using the appropriate PCM.

7.2.4 Medical applications

The storage and transport of medical products is a very suitable application for PCM. The reason is that many medical products are quite expensive and their quality often strongly depends on the storage and transport temperature. Some products need to be transported between 20 °C and 24 °C, others between 2 °C and 6 °C, and others between –30 °C and –26 °C. If possible, the transport vehicles are air-conditioned to the desired temperature, but the transport between the hospital and the transportation vehicle and then between the vehicle and the final destination is still a problem. In the case of blood products for example, it is necessary to avoid freezing and also heating above a certain temperature during transport; otherwise the blood products cannot be used any more. The conventional method to keep the temperature within the allowed temperature range is to use rather complex and expensive cooling systems. Fig.7.12 shows a simple and inexpensive transport system using PCM. The blood transport system, developed by the company delta T, keeps the temperature between 2 °C and 10 °C for 12 hours without additional technical equipment and closes the gaps in the transport chain.

Fig. 7.12. Blood transport system (pictures: delta T Gesellschaft für Medizintechnik mbH) and transport box (picture: AB Aircontainer A.C.).

In well-insulated transport containers, results that are even more impressive are achieved due to the temperature regulating capability of PCM. The Swedish company AB Aircontainer A.C., specialized on transport via planes, gives two impressive examples (Setterwall 2005). The first example is the transport of a medical product from Sweden to Montreal / Canada by airplane in a transport box, as shown on the right in fig.7.12. The medical product has to be kept in the temperature range from +2 °C to +8 °C; it arrived still at +5.6 °C after 105 hours. The second example is the transport of another medical product by airplane from Sweden to Mexico. This product has to be kept at or below -18 °C; it arrived at -18 °C after 56 hours.

7.2.5 Electronic equipment

Most electronic equipment works best in a certain temperature range, and fails if the temperatures are too high or too low. Too high or too low temperatures can also reduce the lifetime of the equipment. For example, telephone-operating companies have a long history of powering equipment with batteries. Current trends are towards remote switches closer to the customer. Because often some electronic equipment must be located and used outdoors, high or low outdoor temperatures ranging from -40 °C to +50 °C can become a problem.

Batteries for example, can show a significant power drop when the temperatures are too low, or complete failure when they get too hot. Further on, the lifetime of a battery is directly related to the load applied, proper recharging, and most of all, the temperature of the battery, which must be kept at optimum conditions. The application of PCM can lead to more constant operating temperature for the batteries, with the PCM allowing peak or cyclic heat loads to be absorbed and then rejected later or at nighttime when ambient conditions allow heat rejection. An example of the application of PCM to batteries is a battery jacket, developed by the company TEAP together with Power Conversion Products and MJM-Engineering. The battery jacket contains bags filled with PCM and is wrapped around the battery. It thereby minimizes the effects of peak heat loads in the day. Information on the design and development of this battery jacket has been published by Cosentino 2000. The danger of overheating is especially high when a battery has to sustain high electrical loads. In that case, the thermal design of a temperature buffer with PCM has to take into account the significant and maybe dominating internal heat gain from battery operation. The description of this application is more complex because significant internal gains occur, and it does not fit into this chapter on applications in transport and storage containers. The interested reader can get more information in the publications by Khateeb et al. 2003, Khateeb et al. 2005, and Mills et al. 2006.

7.3 References

[AB Aircontainer A.C.] AB Aircontainer A.C. Sweden. www.aircontainer.com
[ACME] ACME. Haryana, India. www.acme.in
[Cabeza et al. 2002] Cabeza L. F., Roca J., Nogués M., Zalba B., Marín J.M^a: Transportation and conservation of temperature sensitive materials with phase change materials: state of the art. Presented at 2nd Workshop IEA ECES Annex 17 "Advanced thermal energy storage techniques – Feasibility studies and demonstration projects", Ljubljana, Slovenia, 3-5 April 2002. www.fskab.com/annex17
[Cosentino 2000] Cosentino A. P.: Thermal Management of Telecommunications Batteries using Phase Change Materials (PCM) Jacket^{TM}. IEEE (2000) http://ieeexplore.ieee.org/iel5/7098/19124/00884256.pdf
[delta T Gesellschaft für Medizintechnik mbH] delta T Gesellschaft für Medizintechnik mbH. Giessen, Germany. http://www.deltat.de/english/products/cooling_elements.htm
[Khateeb et al. 2003] Khateeb S.A., Farid M.M., Selman J.R., Al-Hallaj S.: Design and simulation of a lithium-ion battery with a phase change material thermal management system for an electric scooter. Journal of Power Sources **128**, 292–307 (2004)
[Khateeb et al. 2005] Khateeb S.A., Amiruddin S., Farid M., Selman J.R., Al-Hallaj S.: Thermal management of Li-ion battery with phase change material for electric scooters: experimental validation. Journal of Power Sources **142**, 345–353 (2005)
[Métivaud et al. 2005] Métivaud V., Ventolà L., Calvet T., Cuevas-Diarte M.A., Mondieig D.: Temperature controlled packings for the transportation of sensitive products. Presented at 8th Workshop IEA ECES Annex 17 "Advanced thermal energy storage techniques – Feasibility studies and demonstration projects", Kizkalesi, Turkey, 18 - 20 April. www.fskab.com/annex17

[Mills et al. 2006] Mills A., Farid M., Selman J.R., Al-Hallaj S.: Thermal conductivity enhancement of phase change materials using a graphite matrix. Applied Thermal Engineering **26**, 1652–1661 (2006)

[PCM Energy P. Ltd] PCM Energy P. Ltd. Mumbai. Bombay, India. www.pcmenergy.com

[Rubitherm] Rubitherm Technologies GmbH. Berlin, Germany. http://www.rubitherm.com

[Setterwall 2005] Setterwall F.: PCM in insulated containers gives Passive Temperature Control door-to-door. Presented at 8th Workshop IEA ECES Annex 17 "Advanced thermal energy storage techniques – Feasibility studies and demonstration projects", Kizkalesi, Turkey, 18 - 20 April. www.fskab.com/annex17

[Sofrigam] Sofrigam. Nanterre Cedex, France. http://www.sofrigam.com/en/index.php

[va-Q-tec AG] va-Q-tec AG. Würzburg, Germany. http://www.va-q-tec.de

8 Applications for the human body

After discussing applications in transport and storage containers, now the discussion of applications for the human body follows. The reason is that these two fields of application are technically the easiest to understand. Both applications use heat storage usually without internal heat transfer fluid, because their focus is on the temperature control and not on the supply of larger amounts of heat or cold. Applications in transport and storage containers are in most cases basically an insulated environment with a heat capacity increased by PCM. Applications for the human body are quite similar, with two exceptions. First, there is an additional internal heat source, the human body, and second, the insulation by the clothes is usually not as good as the insulation of most transport containers. Both applications are also similar in the way that both are decentralized applications; there is no connection to any energy source. This has made products economic very soon. One of the first applications for example was in space suits for astronauts in the US space program.

The chapter starts with an introduction to the thermodynamic basics of the application for the human body. Then, an outline of different approaches to integrate PCM, for example in clothes, follows. Finally, different application examples are discussed. These examples are far from complete; they just give an overview on typical applications fields and approaches used.

8.1 Basics

The human body produces heat by metabolism. This heat has to be released constantly, while keeping the body temperature in a range suitable for the metabolism at about 37 °C. The heat exchange between the human body and the environment is by thermal conduction, convection, radiation, and evaporation (sweating). If we loose too much heat, the temperature of the body drops below the optimum temperature and we will freeze. If we loose not enough heat, we overheat and start to sweat to increase heat loss by evaporation. The acceptable temperature range for the human body is very small: at 42 °C, only 5 °C above the optimum temperature we die! Therefore, the temperature stabilization by PCM is very promising.

8.1.1 Energy balance of the human body

A range of temperatures, air movement, and relative humidity defines the optimum boundary conditions for heat exchange between the human body and the environment by thermal conduction, convection, radiation, and evaporation. This set of boundary conditions is commonly called *human comfort requirements* and will be discussed in more detail in chapter 9. At this point, to make some crude calculations of the effect of PCM in controlling the temperature of the human body, it is necessary to know only a few facts. These are

- The optimum body core temperature is 36 °C to 37 °C, high enough for metabolic reactions to proceed, and low enough for chemical stability of body substances.
- The metabolic heat production depending on the physical activity is (Çengel 1998) 100 W when resting, 1000 W when working, 2000 W for top athletes in contests.
- The heat exchange at the body surface is by conduction, convection, radiation, and evaporation. This also includes heat transfer from the interior of the body by evaporation when breathing.
- There is a natural temperature control by the heat capacity of the body, which consists to a large fraction of water.

To simplify the following calculations, the effect of evaporation when breathing is neglected. Then, the heat loss at the body surface is

$$\frac{d}{dt}Q = -A_{body} \cdot k_{eff} \cdot (T_{body} - T_{amb}). \tag{8.1}$$

A typical value for the heat transfer coefficient α for free convection in air is about 4 – 8 W/m²K. To take into account the insulating effect of some clothing, the lower value for the effective heat transfer coefficient k_{eff} can be used. Then, as a rough approximation, the human body looses 80 W, assuming a body surface of 2 m² and a temperature difference of 10 K. This is in agreement with the 100 W mentioned above for people at rest.

8.1.2 Potential of PCM

The first way to buffer temperature changes is by the heat capacity of the body. For a person of 70 kg, and taking into account that the body consists mostly of water, the heat capacity is about

$$C_p = 70 kg \cdot 4 kJ/kgK = 280 kJ/K. \tag{8.2}$$

1.5 kg of a PCM with a latent heat of 200 kJ/kg stores 300 kJ, and can therefore delay the cooling or heating of the body by 1 K.

Another way of looking at the situation is to look at the heat fluxes. This can be discussed using the case when a person increases its activity and the additional heat released should be buffered by PCM. If the person releases an additional 100 W due to increased activity, applying 1 kg of a PCM with a melting enthalpy of $\Delta_{pc}h = 200$ kJ/kg will buffer the additional heat load for

$$\frac{m_{PCM} \cdot \Delta_{pc} h}{100W} = \frac{1kg \cdot 200kJ/kg}{100W} = \frac{200kW \cdot s}{100W} = 2000s = 33\min. \qquad (8.3)$$

The same is valid if the body absorbs an additional 100 W of solar heat. This means that, in addition to the passive thermal insulation effect of clothes, clothes with PCM further on provide what is often called active thermal insulation. The active thermal insulation by the PCM controls the heat flux through the clothes and adjusts the heat flux to varying boundary conditions. If the heat generation of the body exceeds the heat release to the environment, the PCM can absorb and store this excess heat. On the other hand, if the heat release exceeds the heat generation, the heat loss can be compensated by heat stored in the PCM.

8.1.3 Methods to apply the PCM

Knowing that PCM can contribute significantly to the stabilization of the body temperature, the question arises how to apply the PCM. The common approach to use PCM to control and stabilize the temperature of the human body is to integrate the PCM into clothes. There are several approaches for this: macroencapsulation, microencapsulation with all its modifications, and composite materials. These are discussed now separately, before the examples, to avoid the impression that for a certain example the approach used for integration is the only one possible.

8.1.3.1 Macroencapsulated PCM

An example of macroencapsulated PCM is pouches. Such pouches can be filled with any kind of PCM, giving flexibility to the PCM selection. Macroencapsulation, with typical dimensions in the range of several cm, offers a cheap way to incorporate large amounts of PCM. The drawback is however that the clothes have to be specially designed to hold the macroencapsulation.

8.1.3.2 Microencapsulated PCM

This drawback does not exist when microencapsulated PCM are used; they can be integrated into textiles already during the production of the textiles. In general, *functional textiles* are textiles with a modified functionality. In the case of PCM for temperature regulation, this means that PCM is integrated into the textiles. Using microencapsulated PCM with a capsule size in the range of several micrometers only, this is possible. With macroscopic pouches, this is not possible, because the pouches cannot be cut to fit an arbitrarily sized jacket etc. At the current state of the technology of microencapsulation, the PCM choice is however restricted to paraffins (section 2.4.3.2). Further on, microencapsulated PCM tend to be more expensive than macroencapsulated PCM. The advantage is however, that textiles with microencapsulated PCM can be used for the fabrication of any kind of clothes, and thus for a more diverse market.

Sariera and Onder 2007 experimentally investigated different methods to manufacture microencapsulated phase change materials suitable for the design of thermally enhanced fabrics and give a series of further references. The combination of microencapsulated paraffin and the fabric can be done in various ways (Hentze et al. 2005, Doshi 2006):

- PCM microcapsules are permanently fixed within the fiber structure. This can be done during the wet spinning procedure of fiber manufacture. Shin et al. 2005 describe the preparation of fibers with PCM and test results.
- PCM microcapsules are embedded in a coating compound. This is done during the finishing process
- PCM microcapsules are mixed into polyurethane foam, and the foam is applied to a fabric in a lamination procedure. This can be done as a layer or as foam dots.

There is a growing number of companies developing and using functional textiles with PCM. The following selection is not a complete coverage; it is only intended to cover the different approaches to integrate microencapsulated PCM.

Outlast® Technologies Inc. / USA uses microencapsulated PCM produced by the companies Ciba and Microtek. The protective polymer shell of the microcapsules is very durable and designed to withstand textile production methods used in fiber, yarn spinning, weaving, knitting, and coating applications. The microencapsulated PCM, called Thermocules™, can be infused into fibers during the manufacturing process or applied as a finishing on fabrics, depending on the application:

- Outlast® Thermocules® in fiber (fig.8.1, left and center). Here, the microcapsules are located inside the fiber. The fibers are spun to yarns, which can then be manufactured to socks, underwear, or knitwear. These are all products worn very near to the skin.
- Outlast® Thermocules® as coating on textiles (fig.8.1, right). Here, different materials can be coated. For example, in sleeping bags nonwovens are coated, in jackets coated linings are used. It can also be applied to midlayers put between the first layer and the lining.

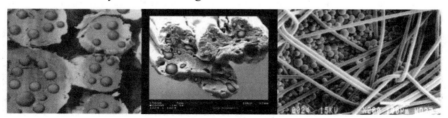

Fig. 8.1. Microscope images of Outlast® acrylic fiber (left), Outlast® viscose fiber (center) and Outlast® coating on textiles (pictures: Outlast Europe).

Schoeller Textil AG / Switzerland developed and produces schoeller®-PCM™, a group of functional textiles using schoeller®-PCM-foam. Fig.8.2 shows the foam applied on a fleece and a special wool mixture with a new schoeller®-PCM™-coating.

Fig. 8.2. Schoeller®-PCM-foam applied on a fleece (left) and wool with schoeller®-PCM™-coating (pictures: Schoeller)

The wool mixture is planned to be applied for winter jackets, coats, and suits. Compared to the millimeter thick schoeller®-PCM™-foam, this coating is significantly thinner and lighter.

8.1.3.3 Composite materials

The third approach to apply PCM is by using composite materials. Currently, the only materials known which are applied in commercial products are Rubitherm powder PX and compound PK. They have already been described in section 2.4.2.1. The powder can be used to fill any arbitrary shaped compartment; the compound can be used to form arbitrary shapes, including sheets, which can be cut to the desired size.

8.2 Examples

There is a large number of companies developing applications for the human body. The following are just a few examples that cover a wide range of applications. The examples do not cover the typical product range of any company. Most companies have a range of products.

8.2.1 Pocket heater

Probably the best-known application of PCM for the thermal comfort of the human body is pocket heaters (fig.8.3). They are used to release heat when a person is freezing. Pocket heaters are one of the few applications where the subcooling of a PCM is intentionally used for an application.

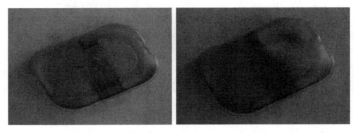

Fig. 8.3. Pocket heater, which uses the effect of subcooling to store heat for a long time and to release it when a person is freezing. Left: subcooled state, right: after crystallization was triggered.

A pocket heater is first heated at home, for example in boiling water, to melt the PCM. Then it cools down to ambient temperature, however without releasing the latent heat. The reason is that a salt hydrate with some additives resulting in strong and stable subcooling is used. Therefore, the PCM in the pocket heater does not crystallize when the pocket heater is cooled to ambient temperature. The subcooling is so stable, that it is possible to carry the pocket heater around for many days without releasing the latent heat unintended. When the heat is needed, a mechanical device integrated into the pocket heater can be used to start crystallization and release the latent heat. The pocket heaters are very convenient for people spending their time outside, like hunters, fishermen, sailors, etc.

In a larger size, such heaters have recently been introduced to the mountain rescue team in Garmisch Partenkirchen / Germany. After an accident, the injured person often lies on the ground without moving. Especially in winter, this can lead to additional health problems. With the newly introduced PCM packs, it is possible to warm up the injured person in a simple and efficient way.

8.2.2 Vests for different applications

An application where macroencapsulated PCM is integrated into clothes is a cooling vest, developed by the company Climator AB. The cooling vest has been developed to cool the body of people who work in hot environments, or with extreme physical exercise. Of course, it is generally also possible to use the PCM to control the temperature of the environment, especially in closed rooms, but many times like in an open desert, this is impossible. In such a case, it is technically easier and more effective to use PCM to control the temperature of the body directly. The vest, shown in fig.8.4, contains compartments to store small pouches filled with PCM.

Fig. 8.4. Cooling vest to cool the body of people who work in hot environments or with extreme physical exercise (pictures: Climator AB)

Heat absorption by the PCM starts as soon as the temperature rises above the melting temperature of the PCM, which is 28 °C. Depending on the application, the duration of the cooling effect is up to 3 hours. This is only possible because of the large amount of PCM, which is about 2.3 kg. If equally distributed across the surface of the vest, the thickness of the PCM layer would be several mm.

The cooling vest is commercially available and is used for many different applications like cooling sports men during breaks in a competition in summer, or to allow fire fighters work longer and closer to a fire. In these cases, it is obvious that the stabilization of the body temperature by a cooling vest is a very good solution. The cooling of the environment is impossible. More detailed information on the vest is available in Ulvengren 2005 and at http://www.climator.com.

A different application of PCM is in ballistic vests, which protect people from gunshots etc. Such a vest is produced by Outlast®.

8.2.3 Clothes and underwear

Applications of microencapsulated PCM in different kinds of clothes are manifold, like in underwear to reduce sweating, based on Outlast® Adaptive Comfort® material, and in many kinds of garments like for motorcycling, including gloves, shoes, etc. Fig.8.5 shows a ski jacket produced by company Colmar using schoeller®-PCM™.

Fig. 8.5. Ski jacket produced by Colmar using schoeller®-PCM™ (picture: Schoeller).

8.2.4 Kidney belt

PCM incorporated in an insulation material will have a temperature regulating effect. This effect is used in a kidney belt for motor cyclists (fig.8.6), which was developed by the company Schoeller textile AG.

Fig. 8.6. Kidney belt with PCM, developed by Schoeller textile AG.

If the temperature of the body or the surrounding increases, the microencapsulated PCM integrated in the kidney belt absorbs the surplus heat. If the temperature falls again, the heat is released.

8.2.5 Plumeaus and sleeping bags

A very common situation where the temperature regulation of the human body is crucial is when sleeping. If the temperature is not adjusted to a comfortable level, we do not sleep well and feel bad the next morning. An active temperature regulation by putting on more clothes when freezing, or undressing when feeling hot, is not a suitable solution because to do this it is necessary to wake up. This is why an automatic temperature regulation, like with PCM, can be advantageous.

There are many companies today selling products in this application field. The integrated PCM stores and releases heat as necessary and thereby buffers the body temperature.

Fig. 8.7. Plumeaus produced by Brennet (picture: Outlast)

Fig.8.7 shows an example where PCM is integrated in plumeaus. Another example is the integration into a sleeping bag.

8.2.6 Shoe inlets

Another point where the temperature regulation of the human body is crucial is at the feet. When the temperature gets too high, we tend to sweat. When the temperature is too low, we will feel very uncomfortable. The company Colortex GmbH has developed a shoe inlet with PCM, based on the compound material Rubitherm PK 31 (section 2.4.2.1) which has a melting temperature of 31 °C.

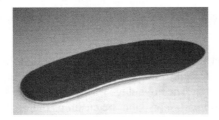

Fig. 8.8. Shoe inlet based on Rubitherm PK (picture: Rubitherm Technologies GmbH)

The shoe inlet has first been presented by Fieback and Lindenberg 2005 and is available commercially nowadays with the brand name REBALLANCE. Other applications of Rubitherm PK are in the textile and leather industry.

8.2.7 Medical applications

PCM can also be applied in medical applications. The use of ice for cooling as a treatment of different sports injuries is state of the art. For some applications, ice is however not suitable. For example, the use of other PCM to lower the body temperature during fever has been suggested. The reason is that common cooling options like cold towels, ice packs, and chilled rooms, often lack a sufficient control of the temperature. For the case of newborns that are in danger of getting brain injuries due to a lack of oxygen, Olsson 2005 has presented first experiments. In these experiments, a salt hydrate with a melting temperature of 28 °C was used to reduce the body temperature, with a young pig replacing the body of the newborn.

Heat therapy is another application field of PCM. The treatment of body parts with heat increases the flow of blood, transport of oxygen, nutritients, waste products, etc. It can reduce pain and support the regeneration of muscles. A straightforward approach to use PCM for heat therapy is to modify the pocket heaters. The company LAVATHERM GmbH followed this approach already more than a decade ago. They developed heat packs in various sizes and shapes, depending on the specific application (http://www.lavatherm.com/frmset1_links.htm). Due to the release of heat over 2-3 hours, the heat treatment not only reaches the skin but also the deeper body parts. Fig.8.9 shows different kinds of hot cushions for heat therapy, developed by the company Rubitherm Technologies GmbH and distributed by the company Fa. Spitzner (Fieback and Lindenberg 2005). The PCM allows a longer application with a constant temperature than conventional methods.

Fig. 8.9. Hot cushions for medical purposes / heat therapy (pictures: Rubitherm Technologies GmbH)

As heat storage material Rubitherm PX powder (section 2.4.2.1) with a melting temperature of 52 °C is used. It consists of an ecologically friendly carrier (silica) which contains a paraffin. The advantage of the powder for this application is that it freely flows in its compartments, and the cushions therefore completely adjust to the surface of the body.

8.3 References

[Çengel 1998] Çengel Y.: Heat transfer – A practical approach. Mc Graw Hill (1998)
[Climator] Climator AB. Sweden. www.climator.com
[Doshi 2006] Doshi G.: PCM in textiles. 2006. http://ezinearticles.com
[Fieback and Lindenberg 2005] Fieback K., Lindenberg G.: New innovative phase change materials of RUBITHERM GmbH. Presented at 8th Workshop IEA ECES Annex 17 "Advanced

thermal energy storage techniques – Feasibility studies and demonstration projects", Kizkalesi, Turkey, 18 - 20 April 2005. www.fskab.com/annex17

[Hentze et al. 2005] Hentze H.-P., Amrhein P., Dyllick-Brenzinger R., Jahns E.: Applications of Micro-encapsulated Phase Change Materials and their Slurries. Proc. of 2nd Conference on Phase Change Material & Slurry: Scientific Conference & Business Forum, Yverdon-les-Bains, Switzerland, 15 – 17 June 2005

[LAVATHERM GmbH] LAVATHERM GmbH. Ingolstadt, Germany. http://www.lavatherm.com

[Olsson 2005] Olsson L.: Usage of PCM in adult, children and neonatal applications. Presented at 8th Workshop IEA ECES Annex 17 "Advanced thermal energy storage techniques – Feasibility studies and demonstration projects", Kizkalesi, Turkey, 18 - 20 April 2005. www.fskab.com/annex17

[Outlast] Outlast Technologies, Inc. Boulder, USA. www.outlast.com

[Rubitherm] Rubitherm Technologies GmbH. Berlin, Germany. www.rubitherm.com

[Sariera and Onder 2007] Sariera N., Onder E.: The manufacture of microencapsulated phase change materials suitable for the design of thermally enhanced fabrics. Thermochimica Acta, vol. 452, Issue 2, 149-160 (2007)

[Schoeller Textil AG] Schoeller Textil AG. Sevelen, Switzerland. www.schoeller-textiles.com

[Shin et al. 2005] Shin Y., Yoo D., Son K.: Development of thermoregulating textile materials with microencapsulated Phase Change Materials (PCM). IV. Performance properties and hand of fabrics treated with PCM microcapsules. Journal of Applied Polymer Science **97**, 910–915 (2005)

[Ulvengren 2005] Ulfvengren R.: Comfort vests. Presented at 8th Workshop IEA ECES Annex 17 "Advanced thermal energy storage techniques – Feasibility studies and demonstration projects", Kizkalesi, Turkey, 18 - 20 April 2005. www.fskab.com/annex17

9 Applications for heating and cooling in buildings

The investigation of PCM for applications for heating and cooling in buildings has a long history. Already in the 1930s, M. Telkes investigated the use of PCM to store solar heat and use it for space heating (Lane et al. 1983). After the oil crisis in 1973, other researchers continued these investigations. However, applications were not yet economic. In the past decade, the economic situation started to change because of rising energy prices. Further on, the focus in R&D shifted from solar heating to cooling. The energy demand to ensure a comfortable environment for humans in buildings has increased worldwide and especially the use of electricity for cooling and air-conditioning is rising fast. Cooling and air-conditioning often cause a peak electricity demand in afternoon hours when, just because of that, peak electricity prices apply. These peak electricity prices and the general trend to higher energy prices have changed the economic situation for the application of PCM in buildings. However, why use PCM for this application? People like to have room temperatures in a very narrow temperature range and a narrow temperature range is exactly the situation where PCM can be used for temperature regulation and for heat or cold storage with high storage density. Especially in buildings with low thermal mass, so called lightweight buildings, the temperature can change quickly and reach lows or highs that make people feel strongly uncomfortable. Therefore, "Applications for heating and cooling in buildings" are expected to have a large market potential for PCM, and in recent years, first products have been successful.

The following discussion is based on several review presentations on this topic, like Mehling 2001 and Mehling 2002, Mehling et al. 2002b, and Mehling et al. 2007. However, the description here is significantly reworked, updated, and extended. The chapter starts with the basics of space heating and cooling. Then, a series of examples for space cooling and for space heating are discussed. Compared to the last two chapters on applications, applications for heating and cooling in buildings are more complicated, and use all different kinds of storage types and a variety of integration concepts.

9.1 Basics of space heating and cooling

9.1.1 Human comfort requirements

People like to have room temperatures in a very narrow temperature range. The reasons for this have been discussed already in section 8.1.1. More general it is however not just the temperature. The heat exchange between the human body and its environment is by different mechanisms: thermal conduction, convection, radiation, and evaporation (sweating). The ambient temperatures, air movement, and relative humidity therefore determine the heat exchange between the human body and the environment. Therefore, a range of each of these parameters defines the optimum boundary conditions for heat exchange between the human body and the environment. This set of boundary conditions is commonly called *human comfort requirements*. The primary function of most buildings is to be a shelter to prevent freezing or overheating of the human body. In other words, the building supplies the human body with an environment that is comfortable.

The human comfort requirements depend on many things, for example clothing, activity, climate, etc, and what is acceptable strongly depends on what is affordable. Therefore, different sources give different values for the range between acceptable and unacceptable conditions. For a general discussion of PCM applications it is however not necessary to go into details, it is sufficient to know typical values. The following values in tab.9.1 for the human comfort requirements are based on data given by Çengel 1998. When discussing applications in a special country, it is advisable to check for the specific regulations.

Table 9.1. Human comfort requirements with respect to heat transfer, based on data given by Çengel 1998

Temperature	
	Air temperature: 22 °C to 26 °C are recommended in moderate climates; in other climates different values can apply.
	Room surface temperatures: changes less than 5 °C vertically and less than 10 °C horizontally are recommended.
Air motion	
	Minimum air motion: necessary for moisture removal at the body surface, especially in summer and in humid climates.
	Maximum air velocity: 9 m/min in winter and 15 m/min in summer is recommended.
	Air exchange rate within a room or building: necessary to ensure sufficient air quality with respect to humidity and dust particles. Values of 0.1/hour for tight buildings to 2 for old buildings have been observed. A minimum of 0.35 is required to meet fresh air requirements for residential buildings and to maintain indoor air quality.

	DIN EN 12831 NA (April 2004) recommends a minimum value of 0.5/hour for standard inhabited rooms and up to 2/hour for meeting rooms.
Humidity	
	Relative humidity: a value between 30 % to 70 % is recommended; about 50 % is considered to be best.

It is important to understand that for the human comfort with respect to heat transfer it is the combination of the temperature of the air and room surfaces, air motion, and humidity, that is important. The reason is that the total heat transfer to or from the human body is determined by the sum over all heat transfer modes: conduction, convection, radiation, and evaporation. For example, with rising air temperature (convective heat transfer), the temperature of the surrounding surfaces (radiative heat transfer) has to be reduced to maintain the heat transfer between the human body and the environment in an optimum for human comfort (fig.9.1).

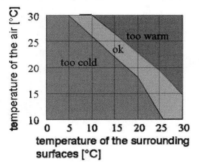

Fig. 9.1. Schematic connection between the influence of the temperature of the air and of the room surfaces on human comfort.

To simplify a discussion, often an *operative temperature* is used instead of giving air and surface temperatures separately.

Human comfort requirements with respect to heat transfer all refer to properties of the air, except wall surface temperatures. The control of these properties in a broad sense is called *air-conditioning*. It includes any form of cooling, heating, and ventilation. Additional requirements to modify the condition of air, however not connected to heat transfer, are cleaning or desinfection of the air. The more common usage of the term air-conditioning is however for cooling and often de-humidification of indoor air. For PCM-technology, the key parameters of human comfort requirements are the temperatures of the air and of the surrounding surfaces, because PCM can only influence these parameters. The relative humidity can be influenced indirectly, because it is a function of the air temperature. All others like air movement, noise level, etc. are boundary conditions for a system. If the application of PCM for heating and cooling of a building is a form of air-conditioning or not therefore depends on what definition is used for the term air-conditioning.

9.1.2 Heat production, transfer, and storage in buildings

To understand how PCM technology can be applied to fulfill human comfort requirements in buildings by heating and cooling, it is necessary to know the main heat fluxes and heat storage mechanisms. The temperature in a building is the result of its heat capacity, heat input and heat losses, and internal heat gains (fig.9.2).

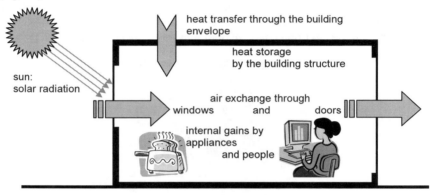

Fig. 9.2. Main influences on the temperature in a building: its heat capacity, heat input and heat loss, and internal heat gains.

Heat storage is mainly by the building structure, at least in solid buildings. Heat input or loss is by transfer through the building envelope, by air exchange through windows and doors, and by solar radiation through the windows. Internal gains come from the heat released by people and appliances, e.g. oven, TV, computer, and of course heating systems. It is not possible to give general values for heating or cooling loads, because these depend strongly on the climate, the building construction, internal gains, and the comfort requirements.

9.1.3 Potential of using PCM

As discussed in section 1.2, applications of PCM can be divided into temperature control and storage of heat or cold with high storage density. In applications for temperature control, the focus is on the temperature regulation and not on the amount of heat supplied. In applications for the storage of heat or cold with high storage density, the situation is the opposite. This is not a strict division; however, it is often easier to understand the basics by discussing simplified cases.

9.1.3.1 Potential of PCM for temperature control

To understand the potential of PCM for temperature control, it is necessary to look at the case without PCM as a reference. Regarding the influence of the building structure on the control of indoor temperatures, there are two extreme cases: tents and caves. Tents have extremely low heat storage capacity; caves are the opposite. In a tent, the temperature can be unbearably high on a summer afternoon and freezing cold during the night of the same day. In caves, the large heat capacity of the cave walls regulates the temperature and fluctuations are often less than 1 K between day and night; in deep caves even between summer and winter. Buildings fill the wide gap between tent and cave. Big churches with massive walls are more like caves and modern lightweight buildings with a wooden frame are more like tents (fig.9.3).

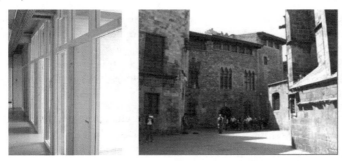

Fig. 9.3. Left: typical lightweight construction with wooden frame and glass windows. Right: typical heavy weight construction with massive stonewalls.

Because of the connection between massive walls and heat capacity, often the term thermal mass is used instead of heat capacity of the building structure. That means heat storage capacity and *thermal mass* are used synonymously; a high thermal mass reduces uncomfortable temperature fluctuations. Many modern buildings have a low thermal mass, because massive walls are missing completely, or because massive walls are only a small fraction of the building structure. Often the interior walls, for example in office buildings, are constructed as lightweight walls because of architectural and cost reasons. Tab.9.2 lists the heat capacities of different building materials and the heat stored in a temperature interval of 4 K, the temperature range given by the comfort requirements in tab.9.1.

Table 9.2. Heat capacities and heat stored in a 4 K interval for different building materials compared to the PCM from fig.3.12 (sample A, 0.5 K/min).

material	c_p per mass [kJ/kgK]	ρ [kg/m³]	c_p per volume [MJ/m³K]	Q/V for $\Delta T = 4$ K [MJ/m³]
EPS	1.2	16	0.02	0.08
mineral wool	0.8	200	0.16	0.64
cork	1.8	150	0.27	1.08
gypsum	0.8	800	0.64	2.56
wood	1.5	700	1.05	4.20
concrete	0.84	1600	1.34	5.38
sandstone	0.7	2300	1.61	6.44
brick	1	1800	1.80	7.20
PCM: peak values	≥ 75	800	≥ 60	
PCM: 22 °C to 26 °C				130

The data show that massive walls made of concrete, sandstone, or brick with volumetric heat capacities of 1.34, 1.61, and 1.8 MJ/m³K are able to store more heat than wood with 1.05 MJ/m³K. Insulation materials like EPS can almost be neglected regarding heat storage.

To understand the potential of PCM for temperature control is somewhat more difficult due to the additional latent heat. Therefore, two different cases of temperature control are used as example:

1. The temperature fluctuates around the phase change temperature more or less evenly, as shown on the left in fig.9.4. In this case, the PCM with a melting temperature at the average temperature generally buffers temperature fluctuations. This however cannot be the general case, because the phase change temperature is fixed, while the room temperatures usually shift somewhat with the seasons.
2. The temperature does not fluctuate evenly around the phase change temperature; it is somewhat higher or lower. Consequently, the PCM slows down the rise or fall of the temperature beyond the phase change temperature, which means it cuts the temperature peaks. This case is shown on the right in fig.9.4.

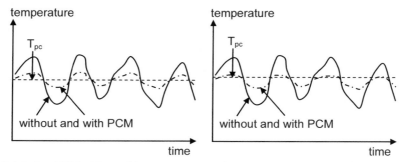

Fig. 9.4. Left: a PCM with a melting temperature at the average temperature buffers temperature fluctuations. Right: a PCM with a higher melting temperature cuts temperature peaks.

To discuss the effect of the PCM quantitatively, it is necessary to take into account that most PCM used in real applications do not have a sharp phase change temperature but a temperature range, at least 1 to 2 K wide. This is in the order of typical temperature fluctuations in a room. Consequently, the melting range has to be considered in calculating the useful heat storage capacity and in the heat flux, which is proportional to the temperature difference between air and PCM. For the case of buffering of temperature fluctuations, fig.9.5 shows the respective heat capacities of materials without phase change, PCM with a phase change temperature, and with a phase change temperature range.

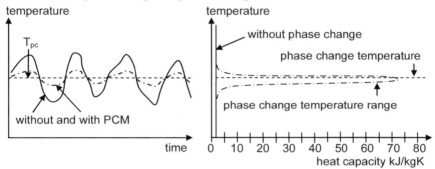

Fig. 9.5. Left: buffering of temperature fluctuations. Right: respective heat capacities of materials without phase change, PCM with a phase change temperature, and with a temperature range.

For a quantitative analysis of the buffering of temperature fluctuations, it is necessary to compare the heat capacity of a PCM with the values for ordinary building materials. An example for the heat capacity as a function of temperature shows fig.3.12 (sample A, 0.5 K/min). In a narrow temperature range, the heat capacity is above 50 kJ/kgK, with peak values higher than 75 kJ/kgK. Assuming a density of 800 kg/m^3, the volumetric heat capacity of the PCM outruns even the materials for massive walls like concrete, sandstone, or brick by more than a factor of 30 (tab.9.2, bottom). The exact value strongly depends on the temperature chosen and changes dramatically within a temperature range of only a few K.

For a quantitative analysis of the second case, where the PCM is used to cut temperature peaks, the situation is different. In this case the PCM is used to avoid that the temperature rises or falls below a certain level (fig.9.4, right), similar as in the application in transport containers. This means it is necessary to know how much heat the PCM can store or release in the temperature interval between the regular or starting temperature and the critical temperature. For an estimate, one can take the width of the comfort temperature range from 22 °C to 26 °C, which is 4 K. In this temperature interval, the PCM stores about 130 MJ/m^3, a factor 18 higher than any conventional material (tab.9.2, bottom). Fig.9.6 compares the different materials in another way: it shows the necessary layer thickness of the different building materials to store as much heat as a 1cm thick layer of PCM. The values are again based on the temperature interval of 4 K.

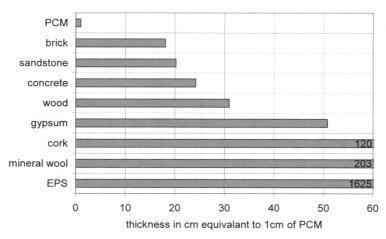

Fig. 9.6. Necessary layer thickness of different building materials to store as much heat as a 1 cm thick layer of PCM. The PCM has 130 MJ/m^3 and the temperature difference is 4 K. Insulation materials are out of scale.

For a massive wall made of brick, a thickness of about 18 cm is necessary, for concrete the thickness is 24 cm. For wood and gypsum boards, as used for lightweight buildings, a wall thickness in the range of 30 to 50 cm would be necessary. Therefore, adding even small amounts of PCM to the building structure can significantly enhance the thermal mass of the building and maybe make a lightweight building perform like a massive building with respect to thermal comfort. The integration of PCM can be as part of separate building components, or as additive to ordinary building materials, which makes them PCM-composite materials. In contrast to ordinary building materials that store heat or cold sensibly, PCM will stabilize the room temperature mainly using its latent heat: when the ambient temperature is below the melting temperature, the PCM releases heat; it stores heat when the temperature is rising above the melting temperature.

Having the necessary thermal mass in a wall does however not automatically mean that it is used. The discussion in chapter 4 showed that it takes a certain time to melt an amount of PCM with a given layer thickness. PCM incorporated into a building in too thick layers will not melt and solidify completely by daily temperature variations. Then, part of the PCM is used only rarely or never and thus is less economical. How much PCM can be used economically? To answer this question it is necessary to calculate the amount of heat that can be stored and withdrawn from a wall. For a heat transfer coefficient from a wall to the air of $\alpha = 8$ W/m^{2}K, typical for free convection in air, the heat transfer resistance at the surface dominates (section 4.1.2) and it is sufficiently accurate to make a quasistatic calculation.

The heat flow into an area A of a wall leads to the melting of a PCM layer with thickness s given by

$$Q = A \cdot \alpha \cdot \Delta T \cdot t = A \cdot \Delta_{pc} h \cdot s \,. \tag{9.1}$$

Therefore

$$s/t = \alpha \cdot \Delta T / \Delta_{pc} h \,. \tag{9.2}$$

Assuming a temperature difference between wall and air of 3 K, a volume fraction of the PCM of 30 %, and a melting enthalpy of $\Delta_{pc}h = 160$ MJ/m³, the result is 1.8 mm/hour. That means, under the boundary conditions assumed here, about 1.8 mm of wall thickness can be used for heat storage per hour, or about 18 mm on a daily cycle with 10 hours.

9.1.3.2 Potential of PCM for heat or cold storage with high storage density

Having controlled the temperature to a constant level does not mean it is at the right level; caves for example have a constant temperature, but usually too low to feel comfortable. This means, besides the thermal mass of the building to smooth temperature fluctuations, it is necessary to have heating and cooling systems to supply extra heat or cold. In such systems, a storage can be used to optimize the performance of the system in case of a fluctuating demand for heat or cold, or in case of a fluctuating supply. The main advantage of latent heat storages is the high storage density in small temperature intervals. In contrast to the application of PCM for temperature control, the phase change temperature or temperature range is now significantly different from the comfortable temperature range (fig.9.7, fig.9.8, and fig.9.10).

In *space cooling applications*, a cold storage can be connected to a chiller, for example. A cold storage allows cold production with the chiller at night, thereby increasing the efficiency of cold production. The storage can also be loaded at low demand levels and support or replace the chiller completely on peak demand. This allows a reduction in chiller size and chiller operation at off peak electricity cost. What kind of storage should be used, latent or sensible? Assuming that cold water at 10 °C is needed for space cooling, with a return temperature of 15 °C, the storage density in a cold-water storage is (15 °C – 10 °C) · 4 kJ/kgK = 20 kJ/kg. This is equivalent to 20 MJ/m³, because the density of water is about 1000 kg/m³. A latent heat storage using a PCM with a melting enthalpy of at least 200 MJ/m³ (fig.2.2) has a 10-times higher storage capacity. The PCM will cause additional cost, but the storage container will be smaller, which saves investment cost and space. Using water-ice as PCM in a latent heat storage reduces the cost of the PCM, however the chiller has to produce temperatures below 0 °C to freeze the ice, instead of only 10 °C needed in the application. This will reduce the chiller performance and increase the running cost for electricity. The best would be to have no chiller at all. In buildings with high thermal mass, nighttime ventilation can be used to store cold of the night in the building structure for cooling in daytime. The temperature difference between room temperature in daytime and air temperature at night outside the building is again only about 10 K. A small latent heat storage can thus replace massive building walls. Such systems, where cold is not produced but taken from a natural source, are called *free cooling systems*. Using natural ice or snow from the winter even *seasonal cold storage* for cooling in summer is possible. The cold and the PCM are for free, and if storage space is available at low cost, seasonal cold storage can be economic.

For *space heating applications*, for example like in solar heating, it is standard to use hot water heat storages with a volume of 0.5 to 1 m³. The temperature difference between the supply and return temperatures in heating systems is in the order of 20 K, however smaller as well as larger values are possible depending on the heat transfer between heating system and room. Assuming that the storage is loaded up to the supply temperature and unloaded to the return temperature, the storage density of a hot water heat storage is 20 K · 4 MJ/m³K = 80 MJ/m³. If the storage is loaded / heated to a temperature much higher than the supply temperature of the heating system the storage density also increases. A latent heat storage could therefore reduce the volume by about a factor of three.

For *domestic hot water* the temperature difference in a heat storage between fully charged and fully discharged is usually much larger than 20 K. Often water is supplied from a pipe in the ground, where it is protected from freezing, at 10 to 15 °C and has to be heated to 50 °C for regular use. Sometimes even temperatures of 60 °C or more are required to kill legionella. With a temperature range of 40 K to 50 K between fully charged and fully discharged however, hot water heat storage already has a storage density of 160 MJ/m³ to 200 MJ/m³. Therefore, the volume reduction using a latent heat storage is about a factor of two or less. Additionally, the necessary heating power to heat domestic hot water from 10 °C to 50 °C is quite high. While it can be supplied easily with a hot water heat storage using direct discharge, and with moderate effort in a hot water heat storage with a heat exchanger on the demand side, it is more complicated to achieve the required heating power in a latent heat storage.

Sometimes *seasonal heat storage* is mentioned as a possible application of PCM because of the high storage density of PCM. Even for low cost PCM with about 0.5 €/kg and a high storage capacity of 0.1 kWh/kg, the PCM cost of 5 €/kWh show that to make a storage economical more than a few dozen cycles are necessary. This makes seasonal storage with PCM, in contrast to seasonal cold storage with snow and ice, uneconomic.

9.1.4 Natural and artificial heat and cold sources

As just discussed, the choice of the right storage for a heating or cooling problem strongly depends on the temperature difference between supply and demand. While the demand side temperatures are determined by the comfort requirements, the source temperatures can vary considerably between different sources.

9.1.4.1 Space cooling

The goal to keep a space cold, more specifically to avoid the temperature rising above a certain level, can be achieved in three ways: the reduction of heat input, the reduction of temperature fluctuations to elevated temperatures, and the improvement of heat rejection. Only for the last approach, a cold source is necessary. Fig.9.7 shows several options of natural cold sources to reject heat from the building interior.

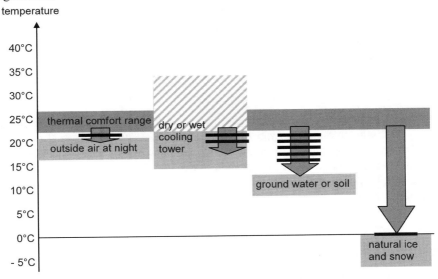

Fig. 9.7. Natural cold sources to reject heat from the building interior. The possible storage temperatures are marked by ▬.

The most common approach to reject heat from a building interior is to use cold night air by ventilation. A disadvantage of this cold source is that it is not absolutely reliable in a hot summer. Another way to reject heat is via dry or evaporative cooling to ambient air. Here the cooling power depends on the humidity and temperature of the ambient air. Also possible is the rejection of heat to ground water and soil. This is usually very reliable, however here the investment cost for the heat transfer can be significant. Last, but not least, natural ice or snow can be used for seasonal cold storage. This way of heat rejection can be useful if natural ice or snow is available in sufficient amounts and with cheap storage space.

Not all cold sources work equally well for all applications. Cold sources are selected with respect to

- cold source temperature

  The lower the temperature of the cold source, the easier the rejection of heat. Further on, a cold storage must operate in the temperature range defined by the comfort temperature and the temperature of the cold source. Therefore, the lower the cold source temperature, the larger the temperature range that can be used by the cold storage.

- availability and reliability

  It is preferable when the cold source is available in daytime, when the demand for cooling is the highest. If outside air at night is used, the storage has to store the cold for use in daytime. However, in hot summer nights the air temperature might not drop low enough. The availability of cold from a wet or a dry cooling tower depends on suitable air conditions; a wet cooling tower does not work well at high humidity and a dry cooling tower not at high temperatures in daytime.

- efficiency (running cost) and investment cost (initial cost)

  Natural cold sources show the highest energy efficiency because the cold is for free or with little electricity consumption. However, investment cost might be high for cooling towers or connections to ground water and soil and, due to small temperature differences between source and demand.

For natural cold sources, there is always a trade off between their advantages in efficiency on the one hand and their availability, reliability, and investment costs on the other. Regarding reliability, artificial cold sources, as shown in fig.9.8, are often preferred. Examples are:

- Electricity driven compression chillers.

  They can produce cold down to temperatures below 0 °C and therefore allow cold storage with artificial ice with high storage density and good reliability.

- Heat driven sorption chillers.

  If the working medium is water, they cannot produce cold too close to 0 °C and other PCM having higher melting temperatures than ice have to be used. The cost for the storage material is then higher as with ice. However, it is possible to increase the chiller efficiency and even solar driven cold production is possible.

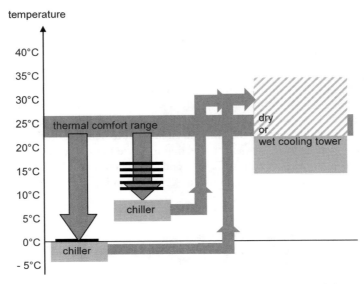

Fig. 9.8. Artificial cold sources: chillers with different temperatures. The possible storage temperatures are marked by ▬.

Artificial cold sources have a lower efficiency than natural cold sources, because they produce cold with an extra energetic effort. This is why they are more reliable; the extra effort makes them more independent from climatic conditions. However, even with natural cold sources the cold is not completely free if a ventilator or a pump is used to transport the cold from its natural source into the building.

Knowing the temperature at the demand side, determined by the comfort requirements, and on the source side, determined by the chosen cold source, the question arises what phase change temperature should be used? The selection of the PCM is a very critical point for space cooling, and due to different climates, building types, and different comfort criteria, there is no single answer. As an example for a discussion, a location in central Europe can be chosen. On the demand side, that is the room, 26 °C are then considered the maximum comfortable temperature. If a temperature difference of 3 K is used to transfer heat from the PCM to air, the upper boundary of the phase change temperature of the PCM is 23 °C. The lower boundary for the phase change temperature is given by the temperature of the cold source, again plus about 3 K for heat transfer. In case of subcooling, the lower boundary determines the lowest temperature when the nucleation starts (fig.2.18). An additional consideration is that condensation has to be avoided, that means in moderate climates temperatures below 18 °C should not be reached at surfaces that are subject to ambient air. This does not mean that the PCM cannot have a lower phase change temperature, but if it does, it is necessary to assure that the cooling surface exposed to the room air does not reach temperatures below the dew point temperature.

In fig.9.7 and fig.9.8, usually several possible phase change temperatures are marked for a LHS. The lower the phase change temperature of the storage material, the higher the cooling power; but on the other hand, the search for a suitable cold source becomes more difficult. For example, the disadvantage of using an ice storage is that the chiller needs to produce cold at lower temperatures and therefore at the cost of a lower COP of the chiller. Even though a detailed analysis of this situation should include the climatic situation, the demand profile, and the options for system layout, it is still interesting to make a rough calculation for three different system designs with different modes of operation. The calculations will show the basic advantages and disadvantages quantitatively.

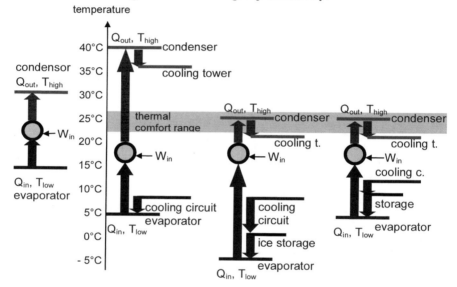

Fig. 9.9. Schematic working scheme of a chiller and three system designs with different modes of operation: in daytime without storage, at nighttime with ice storage, and at nighttime with a LHS with a phase change temperature at 10 °C.

Fig.9.9 shows the working scheme of a chiller and three system designs with different modes of operation: in daytime without storage, at nighttime with ice storage, and at nighttime with a LHS with phase change temperature of 10 °C. The coefficient of performance (COP) of a chiller in an air-conditioning installation is given by eq.9.3, for the ideal, reversible case

$$COP_{AC,rev} = \frac{Q_{in}}{W_{in}} = \frac{Q_{in}}{Q_{out} - Q_{in}} = \frac{T_{low}}{T_{high} - T_{low}} > 1. \tag{9.3}$$

It is a function of the lower temperature, where heat is extracted to cool (evaporator), and the higher temperature, where heat is discarded (condenser). Even for the ideal operation of the chiller, it is necessary to take into account heat transfer at different points, like from the evaporator to the cooling circuit. In the following calculations, a temperature difference of 5 K is used throughout for simplicity.

Table 9.3. Boundary conditions and calculated COP for the different system designs with different modes of operation, as shown in fig.9.9.

	operation in daytime without storage	at nighttime with ice storage	at nighttime with a LHS with phase change temperature 10 °C
outside	35 °C	20 °C	20 °C
condenser	40 °C	25 °C	25 °C
evaporator	5 °C	-5 °C	5 °C
storage	-	0 °C	10 °C
cooling circuit	10 °C	10 °C	15 °C
COP	(273+5)/(40-5) = 7.8	(273-5)/(25-5) = 8.9	(273+5)/(25-5) = 13.9

The results for the COP in tab.9.3 show that using nighttime cold and storage in an ice storage can improve the COP somewhat, and therefore the energetic efficiency of the system is also improved. However, the effect is not very large. The advantage of load leveling with a storage is usually still the dominating advantage because the chiller can be operated at lower running cost using cheap night electricity, because it is not necessary to use expensive peak electricity, and because the chiller can be sized to average instead of peak load which reduces the investment cost. A LHS with a melting temperature of 10 °C however can additionally improve the COP by almost 80 % and thus also the energetic efficiency of the system. However, as a consequence the temperature of the cooling circuit rises which has two negative effects. First, a better heat transfer to the room is necessary to achieve the same cooling power, and second, dehumidification by cooling the air below the dew point temperature is usually not possible anymore.

9.1.4.2 Space heating

The goal to keep a space warm, more specifically to avoid the temperature going below a certain level, can be achieved in three ways: the reduction of heat loss, the supply of heat, and the reduction of temperature fluctuations to too low temperatures. In the last way, a PCM with a phase change temperature (or range) within the comfort range is used to store excess heat from the day from internal sources or solar radiation to prevent temperatures going down at night. To supply heat, an additional heat source at a higher temperature is necessary. Fig.9.10 shows a few examples of natural and artificial heat sources.

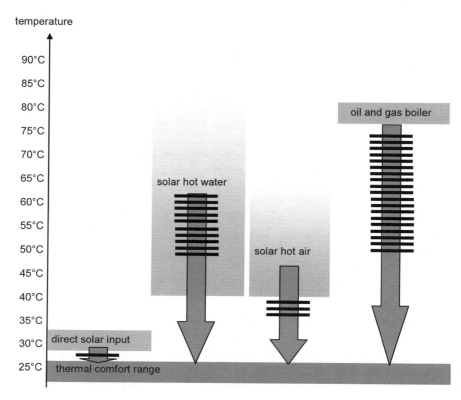

Fig. 9.10. Natural and artificial heat sources. The possible storage temperatures are marked by ▬.

Knowing the temperature at the demand side, determined by the comfort requirements, and on the source side, determined by the chosen heat source, the question arises what phase change temperature should be used? The selection of the PCM for a space heating application is usually less critical than for space cooling. When fossil fuels are used as heat source, the temperature difference between the source and the thermal comfort range is much higher than for cooling applications. The temperature range to select a suitable PCM is therefore broad, but this also means that the competition to hot water heat storage becomes more intense. For solar heating, there is an additional consideration: there is a strong dependence of the efficiency of the collector on the outlet temperature of the collector. The higher the selected phase change temperature, the higher is also the collector outlet temperature necessary to load the storage. This can reduce the collector efficiency significantly.

9.1.5 Heat transfer

While the source and demand temperatures determine the boundaries for the phase change temperature, the mode of heat transfer determines the power and necessary time for loading and unloading the storage. Heat is transferred at several points in a heating or cooling system. At the storage, heat is transferred at least three times: when loading the storage, within the storage material, and when unloading the storage. For the storage alone, the situation was discussed in chapter 5. Storages were divided with respect to heat transfer into

- storages exchanging heat by heat transfer on the storage surface
- storages exchanging heat by heat transfer with internal heat transfer area
- storages exchanging heat by exchanging the storage medium

From top downwards, the heat transfer is improved. It is important not to forget that these options actually apply to the supply side and the demand side separately. For example, the first concept in fig.6.4 shows a storage loaded via a heat exchanger, that is with internal heat transfer area, and is unloaded by heat transfer on the storage surface.

In the case of space heating or cooling within buildings, the final goal is to have a certain room temperature and, if radiative heat transfer is significant, a certain temperature at the heating and cooling surface. This complicates the situation when looking at the system and not only at the storage. On the demand side, that is the room, there is always air as the final heat transfer fluid. Often air that is heated or cooled is transferred to a room, but in many cases heat or cold is also transferred by water first and then to the air in the room via a heat exchanger (fig.6.5, right). Therefore, there are additional boundary conditions in the area of heat transfer on the room side with respect to the heat transfer fluid, heat transfer coefficient, volume flow of the heat transfer fluid, and its temperature. The connection to the room to be cooled or heated is now discussed in more detail.

9.1.5.1 Heating or cooling from a surface

Heating and cooling from a surface works in two ways. The first one is by heating or cooling the air and thereby changing the air temperature. The second one is by direct exchange of infrared radiation between the surface and the human body. As already mentioned in section 9.1.1 on human comfort requirements, a warm surface can give a comfortable feeling even if the air temperature is rather low. This effect is however difficult to calculate and is therefore not discussed here.

When heating or cooling from a surface the heating power is $dQ/dt = A \cdot \alpha \cdot \Delta T$, where A is the surface area, α the heat transfer coefficient, and ΔT the temperature difference between the surface and the ambient air.

Typical values for the heat transfer coefficient α are (according to DIN EN ISO 6946, Oct. 2003)

- for heating or cooling from a wall, which is a vertical surface with heat transfer by radiation and convection: $\alpha = 7.7$ W/m^{2}K
- for heating from the floor or cooling from the ceiling, which are horizontal surfaces with heat transfer upwards by radiation and convection: $\alpha = 10$ W/m^{2}K
- for cooling from the floor or heating by the ceiling, which is a horizontal surface with radiation but with suppressed convection due to the temperature gradient: $\alpha = 5.9$ W/m^{2}K

Typical values for the temperature difference ΔT are

- $\Delta T = 10$ to 20 K for low temperature heating, when using large surfaces
- $\Delta T = 20$ to 40 K for regular heating systems with heating units having smaller surfaces
- $\Delta T \leq 10$ K for cooling, to avoid condensation

9.1.5.2 Heating or cooling by supplying hot or cold air

When heating or cooling by supplying warm or cold air, many things are different. The heating power is $dQ/dt = \rho \cdot c_p \cdot \Delta T \cdot \Delta V/\Delta t$, where $\rho \cdot c_p$ is the volumetric heat capacity of the air, $\Delta V/\Delta t$ its volume flow, and ΔT the temperature difference between room air and supply air temperature. The heat capacity of air $\rho \cdot c_p$ is approximately 1 kJ/m^{3}K and very small compared to that of water, which is approximately $4 \cdot 10^3$ · kJ/m^{3}K. Consequently, large volume flow rates are necessary. These can create noise, or lead to a feeling of uncomfort, if the air moves too fast. However, an advantage of such systems is that heating or cooling and ventilation (air exchange) can be done at the same time. Further on, using air as heat transfer fluid, leakage is no problem compared to systems using water as heat transfer fluid.

9.2 Examples for space cooling

After having discussed all the basics, that is the different requirements for human comfort, the potential of the application of PCM, the different heat and cold sources, and the options for heat transfer, these basics are now applied to discuss different examples. The examples are organized in a logical order, according the approach for heat transfer, with the goal that the concepts are easier to understand. The beginning is examples for space cooling. For the case of cooling from a surface, the PCM can be integrated into building materials or building components. For the integration of the PCM, often an approach as in applications for the human body is followed: microencapsulated PCM is used to integrate PCM into materials and macroencapsulated PCM for building components. The second set of examples is for systems using air as heat transfer medium for charging and discharging. Finally, systems using water or other liquids as heat transfer fluid are discussed.

9.2.1 Building materials

The potential of PCM to reduce temperature fluctuations, and specifically to cut peak temperatures, is quite large, especially in lightweight buildings (section 9.1.3.1). The general concept for cooling with PCM integrated into building materials shows fig.9.11. Heat is stored in the building material in daytime; at night, the stored heat is retrieved from the PCM and discarded by ventilation to the outside.

Fig. 9.11. Concept for cooling with PCM integrated into building materials. Heat is stored in the building material in daytime; at night, the stored heat is discarded by ventilation to the outside.

Since the 1980s, many attempts have been made to incorporate PCM into building materials like plaster, wood, or fiberboards. Usually, the PCM in these attempts was a paraffin or a fatty acid. In the early stage of research, the PCM was integrated by impregnation (Hawes et al. 1993). However, leakage and evaporation of PCM components was a problem because the PCM was only absorbed in pores and channels. The common approach to these problems is encapsulation, but the handling of building materials includes cutting them or putting nails or screws in them. For these problems, microencapsulation is a possible solution. With microencapsulation, there is no loss of PCM under regular use and the amount of PCM released when a capsule is damaged is very small. This means that microencapsulation allows a handing of the building material with PCM the same way as without, including cutting, etc. This is a big advantage as no especially trained craftsmen are necessary for installation. Besides microencapsulation, it is also possible to form a composite between paraffin and a polymer, like HDPE, as described in section 2.4.2.1. According to current literature, this approach also fulfils the requirements for building applications (Yinping et al. 2006).

9.2.1.1 Gypsum plasterboards with microencapsulated paraffin

Gypsum plasterboards are used on a large scale in the construction of lightweight buildings. The incorporation of PCM into gypsum plasterboards therefore has a large potential to increase the thermal mass of such buildings. First investigations to incorporate PCM into gypsum plasterboards date back more than 20 years ago, for example as described in Shapiro et al. 1987. These investigations usually used impregnation to incorporate a PCM. Based on a different approach, the incorporation of microencapsulated paraffin into the plasterboard, the companies Knauf and BASF developed a solution, which assures all the necessary properties for a commercial product, and a suitable production process. These investigations were performed in the course of two R&D projects. Both projects were funded by the German Ministry of Economics (BMWi) between 1998 and 2004: "Innovative PCM-technology" (engl. "Innovative PCM technology") coordinated by the ZAE Bayern and "Mikroverkapselte Latentwärmespeicher" (engl.: "Microencapsulated latent heat storage materials") coordinated by the FhG-ISE. Meanwhile, the gypsum plasterboard with microencapsulated paraffin (fig.9.12) is commercially available from BTC Specialty Chemical Distribution GmbH, a member of the BASF group, with the brand name Micronal® PCM SmartBoard™.

Fig. 9.12. Micronal® PCM SmartBoard™.

The technical data of the PCM plasterboard, and much other information, are available at a special website (www.micronal.de). The standard size plasterboard has dimensions of 2.00 m x 1.25 m x 15 mm and a weight of 11.5 kg/m². The 3 kg/m² of PCM incorporated, which is approximately 26 wt.%, give it a storage capacity of about 330 kJ/m². This gives the 1.5 cm thick plaster board a heat storage capacity comparable to a 9 cm thick concrete wall or a 12 cm thick wall of bricks. Depending on the application and the local climate, the melting temperature of the microencapsulated paraffin can be chosen to be 23 °C or 26 °C. Also important, the handling and installation of the PCM plasterboard is the same as with conventional plasterboard. Information on several reference buildings, where the Micronal® PCM SmartBoard™ has been installed, is available at http://www.micronal.de/portal/basf/ide/dt.jsp?setCursor=1_290222.

At this point, it is interesting to check if the heat is not only stored, but also if it can be discharged on a daily cycle. Assuming for a rough estimate a heat transfer coefficient between wall and air of 7.7 W/m^{2}K (section 9.1.5.1) and a temperature difference of 2 K, the rate of heat transfer is 15.4 W/m^2. Within a time of 7 h for the storage of heat when cooling in daytime, or when discharging at night with colder air, the amount of heat that can be stored is about 15.4 W/m$^2 \cdot$ 7 h = 108 Wh/m^2 = 388 kJ/m^2. Neeper 2000 has done a more detailed investigation. He investigated the thermal dynamics of a PCM wallboard under daily temperature variations to provide guidelines for selecting the PCM and estimating its benefits. His result is that the daily storage capacity is limited to about 300 – 400 kJ/m^2. If more PCM is included, it will not melt and solidify in a daily cycle. Commercial products like Micronal® PCM SmartBoard™ have a PCM content within this limit.

9.2.1.2 Plaster with microencapsulated paraffin

Another option to integrate microencapsulated PCM into a building material used at wall surfaces is the integration into plaster. This option was investigated by the company Maxit and the FhG-ISE within the project "Mikroverkapselte Latentwärmespeicher", already mentioned above for the gypsum plasterboard. They developed a plaster which contains about 20 wt.% Micronal® (section 2.4.3.2), resulting in a heat storage capacity of about 18 kJ/kg in the temperature range from 23 °C to 26 °C. Detailed investigations have been performed in especially designed test rooms at the FHG-ISE: one room with and one without PCM in the plaster. To discard the heat stored in the PCM during daytime, an increased air exchange rate of 5/hour was used at night. The test results of a measurement campaign of several days in the record summer of 2003 show that the plaster with PCM reduced the temperature of the walls by up to 4 °C (Haussmann and Schossig 2006a).

The plaster has been applied and tested in a series of buildings, like the "Dreiliterhaus" – LUWOGE Ludwigshafen (2000), an office in the new building of the FhG-ISE (2002), and the administrative building of maxit in Breisach (2003). A description of several reference buildings contains http://www.micronal.de/portal/basf/ide/dt.jsp?setCursor=1_290222. The plaster is commercially available since 2004 from maxit Deutschland GmbH as "maxit clima®". Information on the product is available at http://www.maxit.de/index1.php3?id=2&schalt=4.

9.2.1.3 Concrete with microencapsulated paraffin

Another option to integrate microencapsulated PCM into a building material is in concrete. Cabeza et al. 2007 present experimental results from a study of two real size concrete test buildings, one of which includes PCM in some walls. The study was performed within the project MOPCON (EU CRAFT ref. G5ST-CT-2002-50331) with the project partners Aspica Constructora (coordinator-Spain), University of Lleida (Spain), Inasmet (Spain), BSA (Spain), Medysys (France), Prokel (Greece), and Intron (The Netherlands). The test buildings, shown in fig.9.13, were installed in the town of Puigverd de Lleida (Spain).

Fig. 9.13. Left: test buildings with walls from regular concrete and from concrete where microencapsulated PCM was added. Right: test buildings with additional trombe wall (pictures: University of Lleida).

The new concrete mixed with microencapsulated PCM is used in the south, west, and roof walls of one of the test rooms. The PCM used is Micronal® from BASF with a melting temperature of 26 °C and a phase change enthalpy of 110 kJ/kg. Each concrete panel contains around 5 wt.% of Micronal®. As a reference, a second test building with standard concrete was built next to it. Both test buildings are fully instrumented to monitor and evaluate the thermal characteristics: temperature sensors in every wall, temperature sensors in the middle of the room at heights of 1.2 m and 2.0 m, and one heat flux sensor in the inside wall of the south panel.

9.2 Examples for space cooling   239

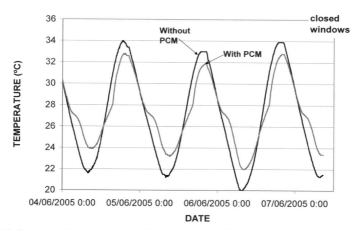

Fig. 9.14. South wall temperatures with and without PCM in June 2005 (picture: University of Lleida).

During summer and autumn 2005, the behavior of the test buildings was investigated. Results were in good agreement with the expected performance. As an example, fig.9.14 shows a comparison of the south wall temperatures of the two buildings. A temperature reduction of up to 4 °C was achieved with the PCM. The PCM included in the concrete walls solidifies and melts in every cycle. The results also showed that night cooling is important to achieve this full cycle every day.

Later, a Trombe wall was installed in the south façade (fig.9.13). To study the effect of the Trombe wall and the PCM it was necessary to find different days with similar climatic data and to compare the measurements of the test buildings with and without PCM.

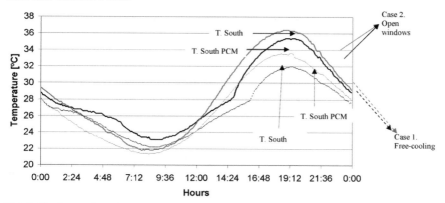

Fig. 9.15. Comparison between south walls with different cases. Case 3 closed windows (9-12/10/2006) and case 2 open windows (27-29/10/2006), both with Trombe wall (picture: Univeristy of Lleida).

Fig.9.15 shows the comparison between having the windows of the South wall open and having them closed. Both cases had similar climatic data with maximum outdoor temperatures of 24 °C and minimum temperatures of 11 °C; the maximum solar radiation was around 700 W/m². Having the windows closed was more effective with temperatures of the walls higher than having them open.

9.2.1.4 Panels with shape-stabilized paraffin

Up to this point, all building materials with PCM used microencapsulated PCM. A second option to integrate PCM into building materials is to use shape-stabilized PCM (Yinping et al. 2006). In 2006, DuPont announced the commercialization of such a building material, DuPont™ Energain®. As fig.2.24 shows, DuPont™ Energain® comes in aluminum-laminated panels that contain a copolymer and paraffin compound. According to DuPont (www.energain.dupont.com), the panels can be applied on the interior walls and ceilings of a building, behind the plasterboard lining, together with a mechanical ventilation system. The panels can be cut, nailed, stapled, or screwed. Once cut, aluminum tape has to be used to seal the edges of the sheets and to cover any holes or abrasions that may occur during installation. The left picture in fig.9.16 shows the installation of a panel.

Fig. 9.16. Left: installation of a DuPont™ Energain® panel. Right: test site for experimental tests in a real building (pictures: DuPont).

The technical data are available at the website given above. The panels measure 1.0 m x 1.2 m and have a thickness of 5.2 mm. Their area weight is 4.5 kg/m². They contain about 60 wt.% of a paraffin that melts and solidifies at around 22 °C and 18 °C respectively. This gives the board a heat storage capacity of more than 70 kJ/kg (= 315 kJ/m²), which is again within the limits calculated before for the amount of heat that can be stored and discharged on a daily cycle.

An experimental test of DuPont™ Energain® in a real building has been performed under the responsibility of DuPont and with the help of the EDF Group (Electricité De France) at a test site in France. The test site is shown at the right in fig.9.16. For the first test phase, it was decided to install the panels in the attic, which is the part of the building most susceptible to overheating, to test the performance of the panels in a worst-case scenario. Therefore, the panels were installed behind the plasterboards, in the walls and in the ceiling of one of the test rooms. A second test room with an identical structure, but without the panels, was used as reference. Both test rooms were separated by a buffer zone.

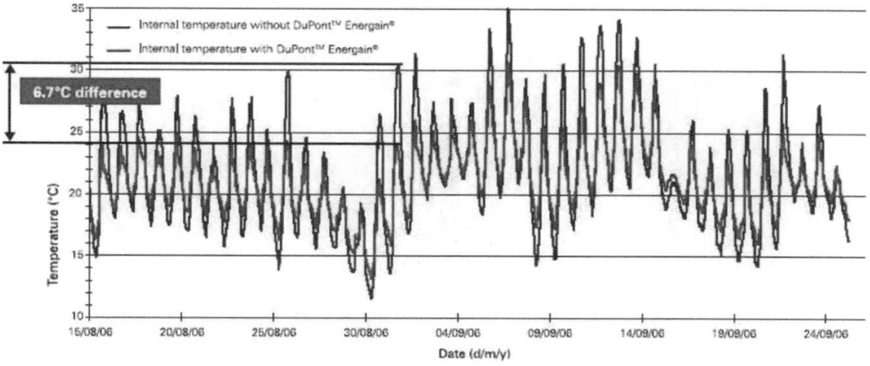

Fig. 9.17. Operative temperatures recorded in the test rooms equipped with DuPont™ Energain® and without (picture: DuPont)

Fig.9.17 shows the operative temperatures recorded in the test rooms equipped with DuPont™ Energain® and without, during the initial six weeks of the experiments in August and September 2006.

A special simulation software program, called CoDyBa, especially adapted for proprietary use with DuPont™ Energain® is available to enable architects and engineers to model with DuPont™ Energain®. The software allows the determination of energy-savings, illustrates responses to temperature fluctuations, and the reductions in CO_2 emissions.

9.2.2 Building components

Besides building materials, also building components can be equipped with PCM. The basic difference is that a component can have a special design and can be fabricated before the building is constructed. Depending on the way of installation, the handling of building components does not necessarily mean that they have to be cut or that nails or screws are put in them. This offers additional possibilities, for example the use of macroencapsulation and salt hydrates.

9.2.2.1 Ceiling with PCM

An example for a building component equipped with macroencapsulated PCM is suspended ceilings. The company Dörken sells a whole range of PCMs under the brand name DELTA®-COOL system. The PCMs are salt hydrates and come in various encapsulations, as shown in section 2.4.3.1. Especially for passive cooling,

Dörken developed a salt hydrate named DELTA®-COOL 24, which melts in the temperature range 22 – 28 °C and crystallizes at 22 °C. It has a melting enthalpy of 158 kJ/kg (= 44 Wh/kg) and a density of 1.6 to 1.5 g/cm^3 for the solid / liquid phase.

Fig. 9.18. Left: Installation of Dörkens DELTA®-COOL 24 on top of a suspended ceiling. Right: Containers equipped with DELTA®-COOL 24 for a field study performed in 2003 (pictures: Dörken)

DELTA®-COOL PCM is packaged in practical aluminum bags of 30 cm x 15 cm. The installation is fast and simple, as the bags can be laid on top of a suspended ceiling, as shown on the left side of fig.9.18. To improve the effectiveness of DELTA®-COOL, materials surrounding the system should provide good heat conductivity and should allow air movement to accelerate the heat exchange. In daytime, when the room temperature rises to 23 °C, the PCM starts to melt and absorbs heat from the room. It thereby stabilizes the temperature around 25 °C in the area close to the PCM, and the room temperature stays at a comfortable level. At night, when the temperature drops below 22 °C, the crystallization process starts, and the previously absorbed and stored heat is released to the surrounding air and the system regenerates.

As a typical value, 8 kg of DELTA®-COOL 24 with a latent heat storage capacity of 350 Wh (1264 kJ) are installed per m^2. According information from Dörken, the ceiling has a cooling power of up to 40 – 45 W/m^2 with natural ventilation. As DELTA®-COOL 24 is a passive system, additional energy supply is not necessary. Dörken recommends combining the passive system with active ventilation for ideal behavior. However, DELTA®-COOL 24 cannot absorb extreme heat loads such as those caused by direct solar irradiation. Dörken recommends that steps should be taken to avoid such extreme exposure.

A field study performed in 2003 on container buildings showed that the temperature was not exceeding 25 °C any more, which is a maximum reduction of 5 °C. Further projects are realized. Additional information is available at http://www.doerken.de/bvf/de/produkte/pcm/index.php.

9.2.2.2 Blinds with PCM

Another example of a building component, which can be equipped with PCM to keep a space cold, is blinds. One of the mayor sources of heat input into a building are solar gains through windows. Especially in modern office buildings, often whole façades are made of glass. To avoid direct solar radiation entering into the rooms behind, blinds can be installed inside the building, or outside in front of the window. Internal blinds cause a thermal problem. As fig.9.19 shows, solar radiation transmitted through the window is absorbed at the surface of internal blinds, they heat up, and release the heat into the room.

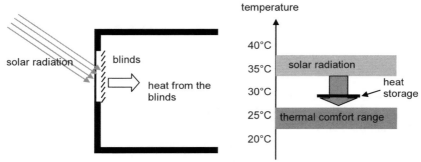

Fig. 9.19. Left: internal blinds absorb solar radiation and release the heat into the room. Right: "space cooling" by reduction of the solar heat input using PCM.

External blinds would avoid this problem, but they have two big disadvantages. First, they are susceptible to strong winds and therefore must be constructed with sufficient mechanical stability; this raises their cost. Second, external blinds are difficult to clean on high-rise buildings. Therefore, in many cases internal blinds are installed. To avoid the problem of heat release from an internal blind into the building it is necessary to reduce the temperature rise of the blinds. This can be done by integrating PCM into the blinds. Then, the blinds still absorb the same amount of solar radiation; however, the PCM integrated into the blinds delays the temperature rise and consequently delays the heat release into the room.

Within the project "Innovative PCM-technology", funded by the German Ministry of Economics (BMWi), the company Warema and the ZAE Bayern have investigated the idea of reducing and delaying the temperature rise of the blinds by integrating PCM. Fig.9.20 shows the prototype of such internal blinds with PCM, which was developed and tested within the project.

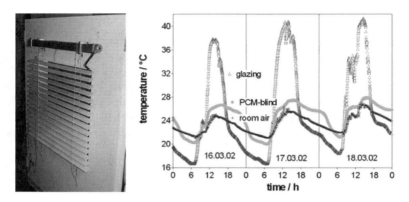

Fig. 9.20. Prototype of horizontal blinds with PCM and experimental results (pictures: ZAE Bayern)

Measurements in a test room under realistic conditions have shown very promising results: the temperature rise of the blinds decreased by about 10 K and was delayed by approximately 3 hours. The air temperature in the room was about 2 K lower. Further investigations by numerical simulation showed a decrease of the operative temperature of the room by about 3 K and a time shift of the heat release from noon to evening. The thermal comfort during working hours is therefore significantly improved. At night, the heat has to be released to the outside by ventilation. Merker et al. 2002 have published more information on this development.

Looking at the heat transfer in this approach to use PCM is quite interesting. Ordinary building materials and components use only free convection for heat transfer to store and to reject heat. In the case of the blinds, comparatively little space and material is necessary because the problem of heat storage is addressed directly at the source. The heat transfer to the PCM is by absorption of solar radiation and therefore not restricted by the heat transfer coefficient for free convection in air. However, in the case of the blinds no internal heat sources are buffered, only the solar heat input.

9.2.3 Active systems using air as heat transfer fluid

The building materials and building components just discussed belong to the category "Storages with heat transfer on the storage surface". In the examples, heat transfer in daytime is by free convection or by radiation, which is passive because there is no actively moved heat transfer fluid. The discharge of the stored heat at night with cold night air is a key issue. The reason is that the difference between the temperature of the air at night and the PCM is usually only a few K. Further on, to cool down the PCM the cold night air has to reach the PCM in the building.

To improve the heat rejection, an air exchange rate of 5/hour like in the study of the plaster with PCM (section 9.2.1.2) can be applied, using an active ventilation system. This assures that the air temperature inside the building is closer to the night air temperature outside. The frequent exchange of air increases the temperature gradient for heat transfer inside the building. However, this does not mean that the heat transfer at the surface of the PCM changes from free to forced convection, because the velocity of the air can still be small.

After the discussion of building materials and components with ventilation of the building to discharge the stored heat at night, a logic continuation is to discuss now systems where the PCM is integrated in a way that the active ventilation leads to a better heat transfer coefficient at the surface of the PCM. For that, the airflow must be directed along the surface of the PCM in a way that heat transfer is by forced convection instead of free convection. Then, it is also possible to use a storage concept with internal heat transfer surface. In the following discussion, successively different aspects of the system design are changed to describe systems that are activated on the demand side, the supply side, or on both sides. This order is chosen in a way that allows the reader to understand the basic ideas behind the different concepts; it does not reflect a continuous improvement of the concepts with the best at the end.

9.2.3.1 Systems integrated into the ceiling

One such concept to use forced convection is based on the integration of the PCM into a ceiling construction. As fig.9.21 shows, the ceiling is constructed in a way that it builds a two-dimensional channel that directs the airflow. The PCM is located in this channel and, in combination with the channel, can be regarded as a heat storage. At night, cold night air is first directed across the PCM surface to discharge the stored heat and cool down the PCM. The air is then discharged to the outside of the building. For cooling in daytime, warm air from the room is forced to move across the PCM, heat is transferred to the PCM, the air is thereby cooled and then supplied back to the room.

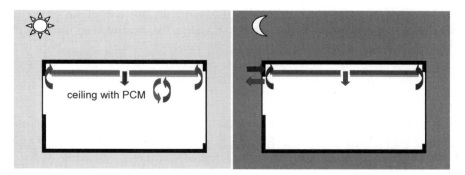

Fig. 9.21. Concept for cooling with PCM integrated into the ceiling. Excess heat from the room is stored in the PCM during the day (left) and discarded at night (right) to the cold night air.

Several modifications of this concept exist. For example when the PCM is in good thermal contact with the suspended ceiling, a second cooling effect is possible because the cold surface of the suspended ceiling will also cool the air in the room by free convection and radiation. The Swedish company Climator uses another modification in a system called "CoolDeck". The system has been installed as part of a demonstration project in the town hall of Stevenage (England). Fig.9.22 shows the ceiling construction.

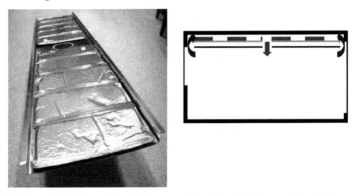

Fig. 9.22. Ceiling with PCM and fan in Climators "CoolDeck" (picture: Climator).

The cool deck consists of the PCM C24, a salt hydrate with a melting temperature of about 24 °C encapsulated in bags, the fan, and a metallic channel to direct the air. As shown on the right of fig.9.22, it is installed directly under the concrete ceiling, and above an already existing suspended ceiling. This way, not only the PCM but also the concrete is used for cold storage. The 5-year demonstration project has started in 2002 and was reviewed in Nov 2005, when about 330 cooling cycles have been performed by the system. Some results and the review report of the demonstration project can be downloaded at the website of Climator. The results show that the maximum room air temperature in summer has been reduced by about 3 to 4 K. The cooling of the building is done without any cold production; the only energy consumption comes from the fan. The ratio of supplied cold to fan energy is in the order of 10 to 20. Economically, the cost of the system is less than of a conventional one. Today the system has already been installed in several buildings.

9.2.3.2 Systems integrated into the wall

It is possible to implement the same approach as for the ceiling in a wall construction, for example if the wall consists of a frame of wooden beams with plates on both sides. This is a typical construction concept for lightweight buildings. Fig.9.23 shows such a system that was built and tested at the ZAE Bayern.

Fig. 9.23. Cooling system integrated into a wall (picture: ZAE Bayern)

The cooling system is installed in the space between the covering plates. Bags filled with PCM are placed on a shelf-like structure and a fan placed at the bottom is used to move the air. Intake and exit of air from the room is via openings at the bottom and the top. At the exit, it is necessary to assure that the volume flow of air, which is necessary to achieve a significant cooling power, does not lead to uncomfortable air velocities. For direct intake of cold night air, optionally an extra intake at the outside of the wall can be used. The tests have not been completed, so no experimental results are available at the moment.

9.2.3.3 Systems integrated into the floor

After discussing the integration into the ceiling and wall, it is not surprising that it is also possible to integrate a similar system into the floor. The general concept of such a system shows fig.9.24. The PCM is located directly under floorboards. During daytime, cooling can be achieved by extracting the warm air from the room, cooling it while melting the PCM, and then bringing the cooled air back to the room. For this, permeable floorboards can be used. At night, cold night air can be circulated under the floor space to cool down the PCM and reject the stored heat.

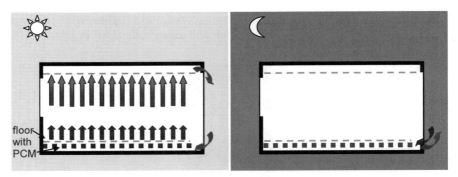

Fig. 9.24. Concept for cooling with PCM integrated into the floor.

A possible modification of this concept is that during cooling, the warm air from the room is discarded to the outside, and replaced by fresh air from the outside that is cooled down by the PCM before it enters the room. In this modification, the system is for cooling and additionally for supplying fresh air by ventilation. Another modification can be made on the discharge of the heat. Instead of using only cold night air, it is possible to add a unit for artificial cooling, for example a compression cooler. In that case, the system is not a free cooling system anymore. Both modifications can as well be used in systems integrated into the ceiling or into the wall, but have not been mentioned previously to allow a discussion of the matter in smaller steps.

An example of such a system is a floor air-conditioning system that has been studied at the University of Hokkaido (Japan). In the Metropolitan areas of Japan, the peak demand for electricity during hot summers is increasing year by year and the daytime load has become twice that of nighttime. The main reason for this is air-conditioning of commercial buildings. Takeda et al. 2006 have focused on a floor supply air-conditioning system in which air is supplied to a room through porous floorboards and a permeable carpet, as shown in fig.9.25.

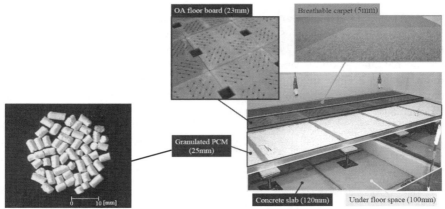

Fig. 9.25. Left: picture of the PCM granulate. Right: construction of the floor air-conditioning system (pictures: University of Hokkaido)

With current storage systems using only the thermal mass of the building, it is difficult to maintain cooling during afternoons. To overcome this problem, the authors have proposed to modify the floor supply air-conditioning system by incorporating PCM. In their system, shown in fig.9.25, latent heat is stored in a PCM that is embedded directly under floorboards in the form of granules. The PCM granules with several mm in diameter form a packed bed, which is permeable to air, so the air can circulate in the floor supply air-conditioning system the same way as without PCM. During night, circulation of cold air through the underfloor space allows to cool down the concrete slab, the floorboard, and the packed bed of PCM. During daytime, the stored cold is used to remove the excess heat from the room. Because of the existing air-conditioning system, it is possible to artificially cool the nighttime air to lower temperatures if necessary to cool down the PCM.

The system has been experimentally investigated in a test room. The PCM used has a melting range from about 18 °C to 21.5 °C and stores 110 kJ/kg. About 12 kg of PCM was used per m^2 of floor area; this is equivalent to 1320 kJ (= 0.37 kWh). Calculations for the test room predict that the complete cooling load can be covered by discharge of cold stored during nighttime without operation of air-conditioners during the daytime, if supply air temperatures in the underfloor space are around 12 to 13 °C. The authors confirmed that the calculation give good agreement with the experiment under similar operating conditions with respect to cold stored during nighttime and the floor and room air temperatures. Takeda et al. 2003, Takeda et al 2005, and Nagano 2007 published more information on the system.

Yamaha and Misaki 1998 investigated the thermal characteristics of flat PCM modules filled with an organic hydrate as PCM for the same application. The PCM has a melting temperature of 13 °C and has been developed by Mitsubishi Chemical Co. Experimental and simulation results for the evaluation of the performance of the storage modules in an air distribution system are described.

9.2.3.4 Decentralized cooling and ventilation unit

Another concept to combine cooling with the option for ventilation is to use a separate unit instead of integration into the building structure as building material or building component. The company Imtech / Germany has developed such a unit since 2004, as part of the Low-Ex project (www.lowex.info) which is funded by the German Ministry of Economics. The decentralized ventilation unit uses a latent heat storage to store cold from night air, which is free, and use it for space cooling in daytime. A schematic drawing of the unit shows fig.9.26.

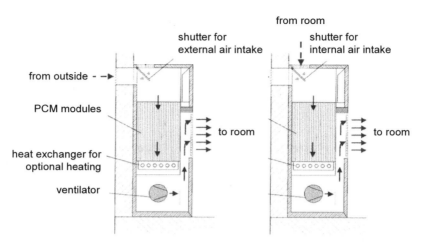

Fig. 9.26. Schematic drawing of the cooling and ventilation unit. Left: cooling down of the unit with cold night air. Right: cooling air from within the room (pictures: Imtech)

As storage material, a paraffin-graphite composite is used to assure sufficient storage capacity and thermal conductivity. It stores about 30 Wh/kg (108 kJ/kg) in the melting range from 18 °C to 22 °C. Each unit is equipped with about 35 kg of storage material and therefore has a storage capacity of about 1 kWh. A stack of parallel plates of the storage material is inserted into the air channel, as shown in fig.9.26. At night, shown on the left in fig.9.26, the system is in the ventilation mode and cools down the PCM and the room with cold night air. A shutter at the top of the unit switches between intake of external air for ventilation and internal air for space cooling. In daytime, shown on the right, the unit draws in warm air from the room, cools it while it passes the storage, and supplies the cooled air back into the room. Optionally, the system can also stay in the ventilation mode and draw in fresh air from outside, cool it at the storage, and supply the cooled fresh air to the room. The air movement is realized by a ventilator at the bottom of the unit, which can move the air at a variable rate. This allows a regulation of the airflow and thereby the cooling power.

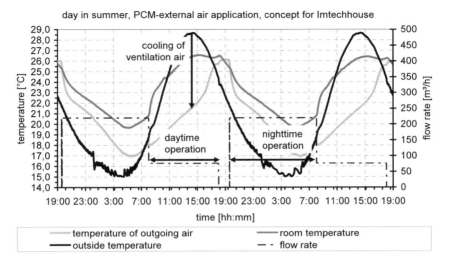

Fig. 9.27. Results of a dynamic measurement in the ventilation mode (picture: Imtech)

First, experiments were performed in a test stand under controlled conditions. Fig.9.27 shows the results of a dynamic measurement in the test stand for conditions typical for a summer day. The unit was tested in the ventilation mode, with the shutter set for intake of outside air. During the day, the temperature of the air at the inlet was rising to a maximum of 28.5 °C, however, when this air entered the room it was cooled down by the storage to about 22 °C, indicated by the vertical arrow. To cool down the storage at night a higher airflow was used.

A set of 50 units has been produced by the company Emco (www.emco-klima.de/de/home.cfm) for tests in the Imtech house in Hamburg. Fig.9.28 shows two of the units with open front pates. The units were installed in a number that 5 kg of storage material was installed per m² of room space, which means one unit per 7 m². The storage capacity per area was therefore 0.14 kWh/m².

Fig. 9.28. Installed units with open and closed front plate (pictures: Imtech).

Building simulations showed that the installed system should be able to keep the operative temperature below 26 °C for almost all times under normal summer conditions. The real tests started in March 2006. Fig.9.29 shows a set of temperature recordings taken on several days in September 2006.

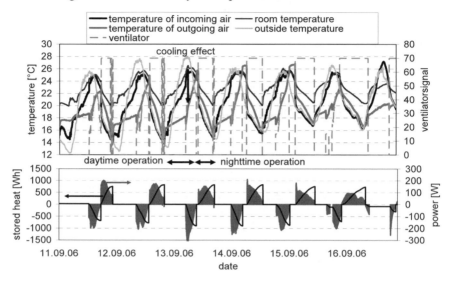

Fig. 9.29. Test results from September 2006 with cloudless weather, measured at a unit installed on the west side of the building (picture: Imtech)

When the ventilator starts to move the air in daytime, the air from the inlet is cooled down by up to 5 K, and then supplied back to the room. The peak cooling power observed on the third day in the graph is about 300 W. The cold stored in the unit was approximately 0.9 kWh, which is consistent with the calculated value above. At night, the storage is cooled down with a higher air exchange rate than in daytime, to assure that the storage is loaded completely. Calculations indicate, that compared to a conventional system using compression chillers, the new system will supply 82 % of the cold using only 5 – 7 % of the electricity (Lüdemann and Detzer 2007). More information on tests of the unit and the real installation is available in the publications by Detzer and Boiting 2004, Lüdemann 2006, and Lüdemann and Detzer 2007. The decentralized ventilation system including control is commercially available since 2007 from the company Emco.

9.2.3.5 Systems integrated into a ventilation channel

The previous sections treated the integration of active systems into the ceiling, the wall, the floor of a room, and finally as separate units. All these integration concepts have one thing in common: they are decentralized. This is an important aspect if different rooms within a building have very different comfort demands, or in large office buildings where different parts belong to different companies that prefer different air-conditioning systems. However, sometimes a building has a central air conditioning system. In these systems, the air is conditioned at one location in the building and then delivered to the different building parts via air ducts.

From the last system described, it is a logic step to try to incorporate the storage directly into the ventilation system of a building. Such a system has been developed by the company Climator with the goal to cool offices using cold from the air at night, at a low price for installations, and with a low energy consumption (Ulvengren 2005). Fig.9.30 shows the system.

Fig. 9.30. Cooling and ventilation system integrated into the ventilation channel, developed by the company Climator (picture: Climator)

A flexible storage design to integrate a storage into the air duct is currently also developed by the company Rubitherm Technologies GmbH (fig.9.31). Their concept uses macroencapsulation in metal sheets as PCM modules (fig.2.30). In order to avoid any fire hazard, different salt hydrates with melting temperatures in the 20 °C to 30 °C range are applied.

Fig. 9.31. Design of the storage currently developed by Rubitherm Technologies GmbH (picture: Rubitherm Technologies GmbH)

A completely different approach to integrate the PCM into an air duct has been investigated by Yamaha 2001. In his study, the PCM was integrated into the coil of a conventional liquid-air heat exchanger, which was placed into the air duct. The PCM was the paraffin n-hexadecane with melting temperature of 18 °C.

9.2.4 Active building materials and components using a liquid heat transfer fluid for heat rejection

All examples discussed until now have used air as heat transfer fluid to discard the stored heat. From free ventilation to forced ventilation, and finally to storages integrated into the ventilation channel, the heat transfer is improved and the rejection of heat is more reliable. Using air as heat transfer medium to discard the stored heat usually means that cold night air is used as cold source. While this is very efficient from an energetic point of view, it is not absolutely reliable that the temperature of the night air drops to a temperature low enough to discard all the heat stored in daytime. The separate unit, and especially the installation in a ventilation channel, can be modified to solve this problem. Because the airflow is controlled in a small chamber or channel, it is possible to integrate a liquid-air heat exchanger and to attach it to a cold source with a liquid heat transfer fluid. In fig.9.26 for example shows a liquid-air heat exchanger for optional heating. It could also be used for additional cooling if cold water is available. Then, the cooling with cold night air can be supported and there is a backup for warm nights. Because then the demand side for cold and the source side have separate heat transfer fluids, it is additionally possible to charge and discharge at the same time. Typical central AC installations already use a liquid to air heat exchanger built into the ventilation channel to cool air with cold water. Such systems can be modified by integrating a LHS into the air channel to give the system additional storage capacity to support cooling at peak demand, or when no cold water is delivered from the cold source.

The integration of a liquid-air heat exchanger into the system has comparatively little influence on the design of the LHS. In this section, the focus is on systems that directly use a liquid heat transfer fluid for heat rejection, usually water or a brine, with the liquid flowing through the storage. Using a liquid as heat transfer fluid for heat rejection allows the use of more reliable natural cold sources, for example natural ground water, or artificial cold sources like compression or absorption chillers, directly (fig.9.7 and fig.9.8). These sources have one thing in common: they deliver the cold with a liquid heat transfer medium.

For the heat transfer on the demand side, that is the room to be cooled, there are two basic options. The first option is that the heat transfer is at a surface within the storage, usually with a ventilator causing forced convection. The second option is that the heat transfer is at the surface of the storage, which is identical with any of the room surfaces, by free convection and radiation. It seems as if up to now only the last option has been investigated experimentally. Such systems are based on the design shown at the top of fig.6.4. Systems without PCM are already well known as *Thermally Activated Building Systems, TABS*. In TABS, building materials or components that are part of the buildings thermal mass are cooled at night in order to shift or reduce peak cooling loads as well as energy consumption (Braun 2003). As source, cold from ground water or soil is commonly used. During the last decade, the thermal activation of building components by such water driven systems has gained an increasing market because it is a promising technology to reduce energy consumption and improve thermal comfort (Stetiu 1999). All designs for TABS have in common, that the liquid heat transfer fluid for heat rejection flows through pipes that are incorporated below the internal surface of a room, or in panels that are suspended from the ceiling. For both approaches, an example with PCM is now discussed.

9.2.4.1 PCM-plaster with capillary sheets

A very common approach for the thermal activation of concrete walls is to integrate capillary sheets as heat exchanger into the wall. Optionally, the capillary sheets can be fixed at the surface of the concrete wall and then covered by a layer of plaster. A straightforward modification is to use a plaster with PCM (section 9.2.1.2). This approach was tested by the FhG-ISE, the BTU-Cottbus, and the companies BASF, maxit, and caparol within the R&D project "Active PCM-storage systems for buildings", funded by the German Ministry of Economics and Labour (2004 to 2007). Fig.9.32 shows how the plaster with microencapsulated PCM is combined with the capillary sheets.

Fig. 9.32. Installation of 1100 m² plaster maxit clima® on capillary sheets (picture: FhG-ISE) in a building in Berlin (picture: M. Schmidt).

For a test, 1100 m² plaster maxit clima® with a thickness of 3 cm were installed on two floors of a building in Berlin. The plaster contains 5 t of Micronal® with a melting temperature of 23 °C. To release the heat, the capillary sheets are connected to ground water. More information on this test is available at www.micronal.de. Haussmann and Schossig 2006b have presented experimental results from a test room. They also discuss the advantages of active cooling. Kalz et. al 2006 outline the effects, the potential, and the performance of TABS in general, with and without integration of PCM. They performed a comparison study between different systems based on a building and plant model in the building simulation software ESP-r. Preliminary experiments in a commonly operated room of a low-energy office building where used to validate the simulation model.

9.2.4.2 Cooling ceiling with PCM-plasterboard

An example where panels are suspended from the ceiling exists also. The company ILKATHERM has developed a plate for dry construction, which consists of a PUR-foam as an insulating layer sandwiched between two coatings made of metal, plastics, plasterboard, wood, or others. It can be used as wall or ceiling element. In this case, the coating is the commercial PCM plasterboard Micronal® PCM SmartBoard™ with a melting temperature of the paraffin of 23 °C (section 9.2.1.1).

Fig. 9.33. Left: Sketch of the cooling ceiling with PCM plasterboard developed by ILKAZELL: 1 metal sheet, 2 PU-foam, 3 capillary tubes, 4 Micronal® PCM SmartBoard™. Right: real installation (pictures: ILKAZELL)

Fig.9.33 shows a sketch of the cooling ceiling and a real installation. As with conventional TABS, different cold sources can be connected to the capillary tubes like ground water or chilled water.

9.2.5 Storages with active heat supply and rejection using a liquid heat transfer fluid

The next step in the discussion of examples for cooling is active cold storages with a liquid heat transfer fluid on both sides. As before, on the cold supply side, where heat is rejected, water can be used to connect the storage to many cold sources. On the cold demand side, where heat is supplied to the storage, using a liquid heat transfer medium allows a better heat transfer within the storage and an easier and energetically more efficient transport of the heat in distribution systems. However, to get finally cold air a liquid-air heat exchanger is necessary. Conventional air-conditioning systems use a chiller to cool a liquid heat transfer fluid that is then used to cool air via a liquid-air heat exchanger. In such systems, a storage with a liquid heat transfer fluid for heat storage and for discharge can easily be installed in the circuit of the liquid heat transfer fluid.

Cold storage is very common in conventional air-conditioning and industrial refrigeration systems. Air-conditioning systems producing artificial cold with electrical or heat driven chillers, rarely operate at full capacity. They usually operate during the day to meet the cooling demand and are designed to satisfy the maximum cooling demand, which occurs only a few days each year. The integration of a cold storage in such a system to supply cold can have different advantages:

- Energetic
    - Reduced peak load of the chiller.
    - Improved energetic efficiency (COP) of the chiller. Usually, the cold demand is highest when the outside temperatures are also the highest. Because the demand side temperatures are fixed by the comfort requirements, higher outside temperatures reduce the COP of a chiller. With a cold storage it is possible to produce cold at times with lower outside temperatures, e.g. at night. This increases the COP of the chiller as long as the temperature necessary to load the storage is not too low (section 9.1.4.1).
    - Chiller operation at operating point with best COP and constant load.

- Economic
    - Reduced energy cost if COP was improved.
    - Reduced investment cost for a smaller chiller, not sized to fit the peak load anymore.
    - Longer operating times of the chiller.
    - Possibility of using low tariff electricity for cold production. This effect can be twofold: cheap off-peak electricity can be used and peak electricity tariffs can be avoided.
    - Discharge of waste heat at night at lower ambient temperatures reduces water consumption for wet cooling towers, or allows the use of dry cooling towers.

- Technical
    - More flexible system operation
    - Back up in cold supply for cases of power failure
    - Reduction of peak electricity loads on the electric grid, if a compression chiller is used
    - Smaller chiller means smaller amount of refrigerant and thus less safety restrictions

Because of its availability, high storage density, and low cost, water-ice is still by far the most widespread PCM. For the cooling of buildings or in industrial cooling processes, the use of ice storages is very common. Due to its high melting enthalpy of about 330 kJ/kg at 0 °C ice storage reaches a much higher storage density than cold-water storage using only the sensible heat. For example, if in an application for space cooling the heat transfer fluid is supplied at a temperature of 15 °C, a cold-water storage can then be operated between lets say 5 °C and 15 °C. The lower limit of 5 °C is to avoid the risk of freezing and, if stratification is used, because water has the maximum density at 4 °C. The cold-water storage will then store 10 K · 4.1 kJ/kgK = 41 kJ/kg and therefore need a 7.5-times larger volume than an ice storage. This is why storages using artificial ice are state of the art, not only for air-conditioning, but also in industrial refrigeration, and as back-up systems. The advantages of using ice storages for cooling in buildings are discussed in detail in McCracken 2004.

There is a wide range of companies producing ice storages and using different storage designs. For example, the company Calmac (http://www.calmac.com) uses a heat exchanger system, and the company Christopia (http://www.cristopia.com) uses a module storage system. In these cases, where the ice is separated from the heat transfer liquid by a wall, the heat transfer liquid must remain liquid at 0 °C and thus brines or water glycol mixtures are usually used. The company BUCO (www.buco-gmbh.de) uses a direct contact system with water as heat transfer fluid. This is possible because a mixing of the storage medium ice and the heat transfer fluid water is not critical; they are the same substance. Last, but not least, there are also systems using ice slurries, for example as developed by the company Axima (www.axima-ref.de). These examples are now discussed in more detail. Where possible, similar systems using other PCM than ice with a melting temperature up to 20 °C are discussed right after the concept is discussed for ice.

9.2.5.1 Heat exchanger and module type storages using artificial ice

Heat exchanger type and module type storages, that is storages with indirect heat exchange, are the most common ones in comfort cooling. The storage type, which is used for the longest time, builds ice directly on the heat exchanger coils of the evaporator of the chiller. For this, the heat exchanger coils carrying the refrigerant are submerged in a tank filled with water and the ice forms on the surface of these coils. To reduce the amount of refrigerant, it is possible to connect the chiller to the storage tank via a secondary circuit carrying a brine. In any case, systems growing ice at the surface of coils are called *ice-on-coil storages*. Because the ice is fixed within the storage, they are also called *static storages*. Two examples from CALMAC and EVAPCO are now discussed.

The basic layout of a CALMAC storage, a heat exchanger type, shows fig.9.34. The storage tank is made of polyethylene, well insulated, and contains a spiral-wound polyethylene-tube heat exchanger submerged in water. Several storages can be combined; therefore, the cold storage system is modular.

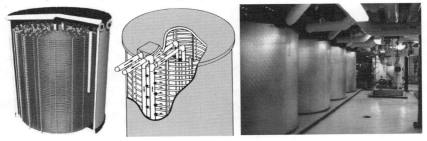

Fig. 9.34. Left: cross section of CALMAC's ICEBANK® cold storage system. Center: flow directions. Right: real installation (pictures: CALMAC).

During charging, a chiller cools the heat transfer fluid (water containing 25 percent ethylene or propylene glycol) to a temperature of several °C below the freezing point of water. It then circulates through the heat exchanger inside the storage and freezes the water surrounding the heat exchanger until approximately 95 % of the water is frozen. For discharging, the heat exchanger is also used. Ice storages with the heat exchanger used for charging and discharging are called ice bank. McCracken 2006 has presented many application examples of this system. Further information, including a list of dozens of examples of installations worldwide, is available at http://www.calmac.com/downloads.

Another ice storage with a typical heat exchanger design was developed by EVAPCO, INC. (USA) and is exclusively marketed by Calmac (http://www.calmac.com/downloads). Fig.9.35 shows a sketch of the system. In this type of storage, cylinders of ice are formed on the surface of steel coils; it is therefore also called ice-on-coil system. Typically, multiple banks of coils are submerged under water in concrete tanks.

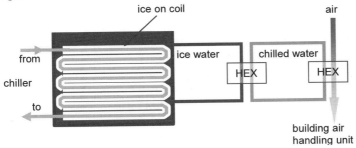

Fig. 9.35. Sketch of a system layout with an ice-on-coil storage.

During the off-peak period, the chiller is operated and cold heat transfer fluid circulates through the tubes of the coils. This causes the water around the coils to freeze. To use the cold stored, that means to melt the ice, there are two options. The first one is that the heat transfer fluid circulating in the coils is used again. The second one, shown in fig.9.35, is that water is circulated over the ice on the coils to extract the cold from the ice by direct contact. The extracted cold is then fed via a heat exchanger into the chilled water loop that supplies cold to the air-handling unit.

The most common module type system is the system developed and marketed by Cristopia. The basic layout shows fig.9.36. The spherical modules consist of a spherical macroencapsulation made from a polymer, filled with the PCM. The picture at the top of fig.9.36 shows a real installation at the French Ministry of Finance, which is operating since 1987.

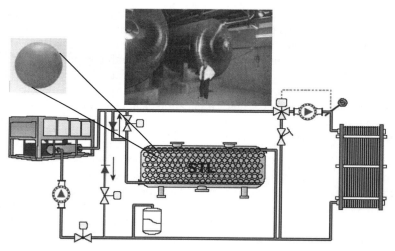

Fig. 9.36. Cold storage based on a module type storage design with spherical encapsulation, as developed by Cristopia / France (pictures: Cristopia).

The concept to integrate the storage into the cooling system, shown in fig.9.36, is with the storage installed in parallel to the chiller. This allows different operation modes: only loading the storage, cooling with the storage only, cooling with the chiller only, cooling with the chiller supported by the storage, and cooling with the chiller while loading the storage (fig.6.3). This is why it is the most often used system configuration. Operating a few valves allows supply and demand of cold with overlap in time as well as storage bypass, that means charging and discharging can be done separately or at the same time.

The following description is only one example of several, which are described at the website of Cristopia at http://www.cristopia.com/english/project/indproject.html. It is an installation at the exhibition center "Fiera Di Rimini" in Italy, which has a total area of 460 000 m^2. The thermal energy storage was completed in 2003. The total load profile of Fiera Di Rimini, shown in fig.9.37, shows that the cold storage reduces the total chiller capacity and shifts the production of cold from on-peak to off-peak electrical period.

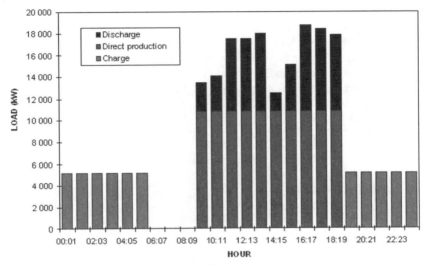

Fig. 9.37. Cooling load profile (picture: Cristopia).

A system layout starts with the simulation or calculation of the load profile of the building in time intervals like 1 h, as shown in fig.9.37. The total daily cold demand has to be supplied by the chiller; in this case, this is about 202 MWh. The maximum cooling demand is 18.8 MW. The plant consists of two centrifugal chillers with a capacity of 2 x 5400 kW and an ice storage installed in series downstream. The storage consists of 4 tanks with a volume of 1200 m³ and a storage capacity of 66 MWh. According to the data in fig.9.37, a single chiller runs from the evening to the morning producing cold only to fill the storage. During daytime, both chillers are running supported by the storage. The chiller capacity of 2 x 5400 kW is only 57 % of the peak cooling demand; the storage supplies the missing 43 % and thereby shifts 18.4 MWh$_{el}$ from on-peak to off-peak electricity period. It thereby saves operating costs, maintenance, and demand charge.

From the discussion in section 5.4.3 it is also possible to understand to some degree the technical properties of the storage system with respect to storage capacity and power. The phase change enthalpy of water-ice per volume is 330 kJ/kg · 0.92 kg/m³ = 304 kJ/m³ = 0.084 MWh/m³. The storage volume is 1200 m³, this means it would store 1200 m³ · 0.084 MWh/m³ = 101 MWh if the storage contains ice to 100 vol.%. According to the data given by Cristopia, the real capacity is 66 MWh, which is about 64 %. This can be understood from the maximum volume fraction for the spherical modules of 74 vol.%, which does not yet take into account the volume occupied by the capsule wall and the air enclosed. For an estimate of the power of the storage, it is necessary to calculate the heat transfer surface in a first step. The volume of the modules is about 64 vol.% of 1200 m³ that is 768 m³. A single nodule has a radius of about 0.05 m, which means a volume of 0.00039 m³. This means the number of modules is about 2 million. Their surface area is 0.01 m², and their total surface area is therefore 20000 m². At the end of section 5.4.3.3 it was discussed that under some assumptions the power of a module type storage can be calculated as if it were a heat exchanger type storage. Using the secondary leaving temperature in the system of +6 °C, and the secondary return temperature of +13 °C, the respective solution in eq.5.10 leads to

$$\dot{Q} = A \cdot k \cdot \frac{\Delta T_{in} - \Delta T_{out}}{\ln \frac{\Delta T_{in}}{\Delta T_{out}}}$$

$$= 20000 m^2 \cdot k \cdot \frac{13K - 6K}{\ln \frac{13K}{6K}}$$

$$= k \cdot 181068 m^2 K \qquad (9.4)$$

The peak power supplied by the storage is 43 % of 18.8 MW, which is 8.1 MW. Then, the heat transfer coefficient k would be k = 8.1 MW / 181068 m²K = 45 W/m²K. According to tab.4.1, this is at the lower end of the values for forced convection in liquids. This is not surprising for two reasons: first, the thermal resistance in the PCM and the capsule wall is neglected; second, due to the large volume fraction of the heat transfer fluid in the module type storage the flow velocity of the heat transfer fluid is probably rather slow.

More examples of installations are described on the website of Cristopia. This includes also an installation in a district cooling plant in Bangsar (Malaysia), where the storage size is 1900 m³. Velraj et al. 2002 published a detailed analysis of an installation at Tidel Park, a software office complex with 12 storeys and a building carpet area of about 92900 m², in Chennai / India. The authors discuss the reason for using this storage system in an air-conditioning application, its size, the advantages of using combined sensible and latent heat storage, and the selection of an optimized nodule diameter. Further on, the possibility of using this concept in other air-conditioning applications like hotels, office buildings, and residential complexes is discussed. Hiroshi 1998 did a further interesting publication. He describes a system with an ice storage installed by Mitsubishi Chemical Engineering Co., who has a license from Cristopia for the Japanese market, in a project done for Minato-mirai 21 District Heating and Cooling Co. The installation was completed in 1994. The author describes the system design and operation records very detailed. Readers interested in more technical details are referred to the book by Dincer and Rosen 2002, which also describes several application examples and discusses them.

9.2.5.2 Heat exchanger and module type storages using other PCM than ice

ASHRAE 1991 mentions an application using a PCM with a phase change temperature of 8 °C and a phase change enthalpy of 95 kJ/kg, which is based on Glaubersalt. For encapsulation, containers with walls from HDPE were used. The system was charged with water at 5 °C, and was able to supply cold water at 9 °C. This information is however more than 15 years old. Currently it seems as if there is little activity to develop storages using a heat exchanger or module design with PCM having a phase change temperature in the range suitable for space cooling. Compared to water/ice, a significant disadvantage is the higher cost of the storage material. On the other hand, the higher phase change temperature promises a higher energetic efficiency in cold production. This promise however could be hard to fulfill, taking into account the necessary temperature difference between PCM and heat transfer fluids in these storage types.

9.2.5.3 Direct contact type storage using artificial ice

If ice is used as storage medium and if the demand cycle of the application uses water as heat transfer fluid, at least a direct contact heat transfer on discharging the storage is possible. This option was already discussed with the storage designed by EVAPCO: the ice is generated on a heat exchanger located in the storage tank, and discharge can be by direct contact between the ice and the heat transfer fluid water. The available surface for direct contact heat transfer is however limited, and would become zero if the water in the tank were frozen completely. To avoid these problems, a more common approach is to generate the ice outside of the storage, store it as smaller ice sheets in the storage tank, and discharge the cold by direct contact of water to the ice. In this approach, the contact surface between ice and water is much larger and can never become zero. Such systems, consisting of an ice producing section and an ice/water storage section, are commonly called *ice-harvesting systems* (fig.9.38). The ice producing section, also called *icemaker*, *ice harvester*, or *ice generator*, consists of the cold source, usually a compression chiller, and a flat or cylindrical evaporator where the ice is formed at the outer surface.

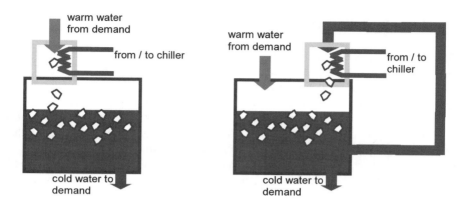

Fig. 9.38. Ice harvesting system with two integration options for the icemaker: left, connected to the warm water return; right, integrated in a closed loop of the storage.

For charging the storage, water is pumped from the demand (fig.9.38, left) or from the storage tank (fig.9.38, right) to the distribution system above the icemaker. There it is distributed, and then flows down the outside of the heat exchanger of the chiller's evaporator as a falling-film. During this process, the water is cooled and a fraction of it forms ice on the surface of the heat exchanger. After a certain time, ice sheets with a thickness of 5 - 10 mm cover the surface of the heat exchanger. To release (harvest) the ice, hot gas is used for intermittent heating. The released ice sheets then fall into the storage tank below, which is partly filled with water. This finishes the charging process. For discharging, water with a temperature above 0 °C is pumped into the storage and then cooled by direct contact with the ice sheets. Because of the large surface of the ice sheets, it is possible to reach water temperatures as low as 0.5 °C. The cold water is then separated from the ice employing the density difference and pumped to the demand.

Such a system to produce cold water has for example been developed by the company BUCO Wärmeaustauscher GmbH (www.buco-gmbh.de). While they are common in industrial cooling, there seems to be no example where they have been applied space cooling.

9.2.5.4 Storages using natural ice and snow

In recent years, similar systems are however applied when the ice is not artificial but coming from a natural source. Applications of natural ice and snow for cooling started with applications to preserve food thousands of years ago. Already Stone Age people have used ice caves to preserve food. In the Antique, the Romans used ice and snow from the Apennine Mountains to cool food and drinks. Until the 19th century, it was common practice to produce ice blocks in the winter, store them in cool houses, and sell the ice in summer. Because the use of natural cold sources like ice and snow was inconvenient and not reliable, attempts to produce artificial ice started. From around 1900, machines to produce ice became available commercially and the use of natural ice disappeared. In recent years, a Renaissance of the use of natural ice and snow can be observed. The reason is that even in snow rich mountain areas the demand for space cooling in the summer months becomes significant.

Seasonal snow or ice storage actually comprises two sources: first, natural ice and snow from precipitation, for example from roads, and second, snow or ice produced artificially using natural cold like cold air. In both cases, the energy for cooling is practically free. The storage vessel employed is usually a more or less watertight pond, which is located indoors, outdoors on the ground, or underground. The heat transfer fluid is air, melt water, or ground water. To extract the cold, the heat transfer fluid flows through the snow in direct contact and melts the snow while being cooled down. The cooling power is controlled by varying the flow rate. After use, the heat transfer medium is rejected or recirculated. According to Skogsberg and Nordell 2006, who thoroughly discuss the technology and boundary conditions for snow cooling, there are a number of suggested and implemented techniques of snow and ice storage for cooling applications. In Japan, about 100 projects were realized during the last 30 years, and in China, there are about 50 - 100 snow and ice storage systems. Canada, the USA, and Sweden also have made efforts to apply this technology.

A well-documented example is a large snow deposit installation for cooling a regional hospital with 190000 m² floor area in Sundsvall / Sweden (Skogsberg and Nordell 2006). Fig.9.39 shows an outline of the Sundsvall snow cooling plant.

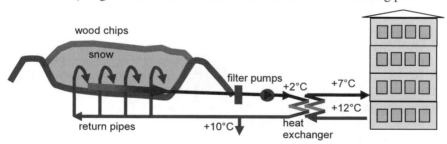

Fig. 9.39. Outline of the Sundsvall snow cooling plant.

The storage is an open, shallow, watertight pond with an area of 130 m x 64 m. It is designed to fit 60000 m³ of snow, which are thermally insulated by woodchips. The first cooling operation started in June 2000. Depending on the snowfall, snow guns artificially produce almost half of the snow. More than 75 % of the total cooling load was delivered by the snow system over the years. A further expansion is planned in about 2010. More details on the construction and operation experience have been published by Skogsberg and Nordell 2006 and Nordell and Skogsberg 2007.

Readers interested in more general information might look at Skogsberg 2005, who gives a comprehensive review of different snow storage systems and applications in his PhD thesis "Seasonal snow storage for cooling applications". Regarding systems using air as heat transfer fluid, a short discussion can be found in Mehling et al. 2007.

9.2.5.5 Direct contact systems using other PCM

Direct contact systems have also been investigated with paraffins and salt hydrates as PCM. To use the method of direct contact heat transfer, such systems use a heat transfer medium that does not mix with the storage material. In the case of salt hydrates, this is usually some kind of oil. The list of salt hydrates that have been investigated for their applicability in direct contact systems (section 5.4.2.1) however does not include any material with a melting temperature suitable for cooling in buildings. This is not too surprising as section 2.2.2 lists only two salt hydrates in the 0 °C – 20 °C range, and both have significant disadvantages. In the case of paraffins, the situation is different. There are several examples of paraffins and mixtures of paraffins melting in the 0 °C – 20 °C range (section 2.2.2). Investigations have been performed that include the search for suitable paraffins and mixtures, as well as experimental investigations in a lab scale storage with paraffin RT5 as PCM and water as heat transfer fluid (He et al. 1999, He and Setterwall 2002). The system was investigated for space cooling including the installation in district cooling systems, and an economic analysis was performed.

9.2.5.6 Slurry type storages using artificial ice

The interest in phase change slurries (PCS) for applications in space cooling originates from its dual use as storage medium and as heat transfer medium. As storage medium, the fraction undergoing a phase change can help to achieve a higher storage density compared to storages using only cold water. As heat transfer medium, it transports more cold with the same volume, and it leads to higher heat transfer coefficients. From these primary advantages follow secondary advantages when designing a new cooling system. On the side of investment cost, these are the use of smaller storages, the use of smaller chillers because of peak shaving, the use of smaller heat exchangers because of better heat transfer, and the use of smaller piping equipment etc. because a smaller volume flow is sufficient to transport a certain amount of cold. On the side of energy savings and reduction of running cost, there is reduced energy consumption for pumping, reduced heat losses due to isothermal storage, and a smaller temperature lift. In the case of retrofitting an existing installation, the performance can also significantly be improved: just by exchanging the heat transfer fluid, the storage and transport capacity can be improved. Even in systems without storage containers, using a phase change slurry as heat transfer fluid can be a significant advantage. If the evaporator of the chiller is directly connected to the cooling demand, this can lead to large amounts of refrigerant in the pipe system. When using NH_3 as refrigerant this can be a safety problem, when using CFCs it is an environmental problem when. To reduce the amount of refrigerant, the primary circuit using the refrigerant is often restricted to the production of cold. The cold transportation between chiller circuit and demand is then by a secondary circuit. The secondary circuit can be filled with an ordinary heat transfer fluid, however when a phase change slurry is used instead there is an advantage: the solid-liquid phase change allows similar heat transfer rates and heat transfer coefficients like the liquid-gas phase change of the primary refrigerant. In such systems, the phase change slurry is then called a *secondary refrigerant*.

The most common slurries are ice slurries. The production of an ice slurry can be done using similar ice maker equipment as described in section 9.2.5.3. However, in contrast to the production of ice sheets, in the case of an ice slurry fine crystals with a diameter of less that 1 mm have to be produced to assure that the slurry can be pumped. This can be achieved using any traditional brine, like water-salt or water-glycol solutions. Fig.9.40 schematically shows the solubility curve of a water-NaCl brine.

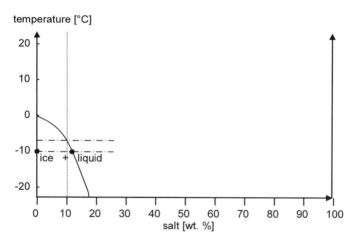

Fig. 9.40. Schematic solubility curve of water-NaCl. When a mixture with 10 wt.% is cooled to about -7 °C ice starts to form.

When a mixture with 10 wt.% NaCl is cooled to about -7 °C the first ice starts to form and the salt concentration in the remaining liquid increases. Generally, ice slurries can be applied at any temperature in the range from -2 °C to -50 °C by choosing the suitable brine; however, in space cooling the lower temperatures result in a reduction of the chiller efficiency. According to ASHRAE 1991, the use of the brine in the icemaker will automatically lead to a formation of small ice particles. Most other literature and commercial suppliers however additionally mention the use of some kind of a mechanical scraper. Fig.9.41 shows the most common ice making technologies where ice is grown at the surface of a heat exchanger. They are classified by the approach to release the ice from the heat exchanger. In the case of ice sheets, which are produced from cooling pure water at the outer surface of cylindrical (pipe) or plate heat exchangers, intermediate heating is used to release the ice sheets, as already discussed above.

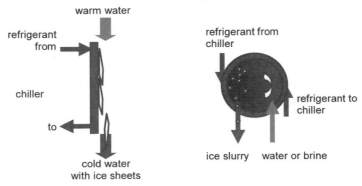

Fig. 9.41. Overview of the most common ice making technologies where ice is grown at the surface of a heat exchanger: left, production of ice sheets; right, production of ice slurries.

In the case of ice slurries, usually produced from cooling a brine, different kinds of mechanical scrapers like rotating knives, rotating cylindrical slabs, rotating brushes, or screws are used to remove the ice. For this, the heat exchanger usually has a cylindrical shape, and the ice is formed at the inner surface of the cylinder where it can be removed by the rotating mechanical scrapers. For the production of an ice slurry, there are also other methods of ice making where ice is not grown at the surface of a heat exchanger to avoid the problem of removing the ice from the heat exchanger surface. Fig.9.42 shows an example: subcooled water is produced inside a heat exchanger, and then ejected against a plate to cause nucleation.

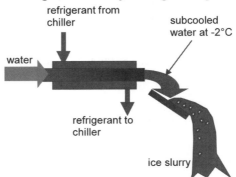

Fig. 9.42. Production of an ice slurry from subcooled water: nucleation is caused by collision of the subcooled water with a plate.

Readers interested in more information about ice making should look at the publications by Dincer and Rosen 2002, Egolf et al. 2003, or Sari et al.

Ice slurries are by far the most common kind of slurry for cooling applications, especially for industrial process cold like for the cooling of food. A commercial example of ice slurries is the deepchill™-system. The deepchill™-system, developed and patented by Sunwell Technologies Inc. in Canada, uses ice crystals in a carrier fluid. Axima Refrigeration markets this technology in Europe. More information on the system is available at www.axima-ref.de and http://www.sunwell.com. A short description of an installation for cooling in a building is available at http://www.sunwell.com/suntech12.htm. It is an installation at the Ritz Carlton Plaza of Osaka, installed in February 1997. The hotel has forty floors and is equipped with 31 ice generators, each having a capacity of 260 kW, and 16 sets of ice storage tanks with 140 or 70 m^3 and a total capacity of 80750 kWh. With the storages, an annual cooling shift rate as high as 81 % and a reduction of the energy consumption for pumps and air-conditioning from 12.5 W/m^2 with conventional air conditioning to 8 W/m^2 is achieved. Another very large system using ice slurry for cooling in buildings is installed in the station building in Kyoto / Japan (Mehling et al. 2007). The ice slurry tank has a size of 1700 m^3.

9.2.5.7 Slurry type storages using other PCM than water / ice

The generation of an ice slurry for space cooling is not the best solution regarding the energetic efficiency of the system. The reason is that the phase change temperature is much lower than the temperature that is needed for space cooling; this reduces the chiller efficiency. For this reason, there is significant R&D to find a suitable PCM for a slurry to store cold at temperatures between 5 °C and 18 °C. Mainly three approaches have been investigated for this application:

- Slurry based on microencapsulated PCM; in this case the PCM is usually a paraffin
- Slurry based on an emulsion; in this case the PCM is usually a paraffin
- Slurry based on a clathrate

Probably the best-investigated approach is slurries based on microencapsulated PCM, that is more specific a carrier fluid, for example water or a water / glycol mixture, and microencapsulated paraffin within a shell usually made of PMMA. The size of the capsules ranges between 1 and 20 µm. Gschwander and Schossig 2006 have investigated slurries of microencapsulated paraffin in water (fig.9.43).

Fig. 9.43. Slurry based on microencapsulated paraffin (picture: FhG-ISE).

Because of the microencapsulation, these slurries can be pumped no matter if the paraffin is solid or liquid. A fraction of up to 40 % of microencapsulated paraffin has been tested; however due to the high viscosity most experiments were performed with a fraction of 30 %. The paraffins tested were paraffin C14 (melting temperature 6 °C) and C16 (melting temperature 18 °C). Enthalpies in the range of 60 kJ/kg to 70 kJ/kg were achieved; this means about four times as much heat or cold is stored compared to pure water in a 4 K temperature range. However, experiments showed that both paraffins subcool down to 0 °C and 11 °C respectively, so that a temperature range somewhat larger than 4 K has to be used. Gschwander and Schossig 2006 could also confirm experimentally for the materials and conditions used that the ratio of transported heat to the pump energy is better for the slurry than for water. This is despite the higher viscosity of the slurry compared to water, because the slurry transports more heat per fluid volume. A potential problem of microencapsulated PCM slurries is still the possibility of destruction of the encapsulation due to shear forces in pumps etc. This would release the paraffin, and consequently the paraffin could separate and block channels or pipes. This can be avoided when the slurry is just used within a storage container and heat input and output is via a heat exchanger. The drawback of this is however that many advantages discussed above are lost. Pollerberg et al. 2005 have also investigated slurries for cold supply based on paraffin / water emulsions and of microencapsulated paraffin / water. The paraffin chosen was tetradecane because of its melting temperature of 5 °C. They discuss the thermal and rheological characteristics of the slurries, as well as environmental aspects, the technological feasibility, and economic considerations for the future use of these kinds of slurries in cold supply systems.

Clathrate slurries have also been investigated for use in space cooling. Darbouret et al. 2005 have investigated Tetra-n-Butyl Ammonium Bromide (TBAB) aqueous solutions, which crystallize into hydrate slurries under atmospheric pressure condition and temperatures between 0°C and +12°C.

Even though that a variety of different slurries besides ice-slurry has been investigated for their use in cooling applications, it seems as if only for the case of the microencapsulated paraffin slurry there is a large scale application: it is a cold storage for cooling in Narita Airport in Tokyo / Japan. A detailed description of the system was presented by Shibutani 2002 and is summarized now. The problem at the installation at Narita airport was that the refrigerant in the existing chiller had to be substituted by a different one, due to environmental reasons. This reduced the capacity and performance of the existing chiller. In order to allow the chiller with reduced capacity and performance to fulfill the existing cold demand of the whole system, the storage was installed. The demand peaks occur between 8:00 and 22:00, so cold produced at night by the chiller can be stored and used at peak load in daytime. The development of the new storage and the installation of the system were in cooperation between New Tokyo International Airport Authority and Mitsubishi Heavy Industries Ltd., with advice by Tokyo Electric Power Company. The characteristic temperatures of the demand side are a supply temperature of 5 °C and a return temperature of 12 °C. Therefore, a microencapsulated paraffin slurry with phase change between 5 °C and 8 °C was selected. The average capsule diameter was about 2 µm. Durability tests were performed for 25 years of operation by accelerated thermal stress and mechanical stress tests. Within the system, the storage with 970 m^3 of slurry is installed in parallel to the chiller and separated from the remaining system by a plate heat exchanger. The slurry is pumped by a centrifugal pump out of the storage, to this heat exchanger, and back to the storage. According to Shibutani 2002, the slurry has a storage density of 67 MJ/m^3, less compared to 167 MJ/m^3 for ice storage, but significantly more compared to 21 MJ/m^3 for cold-water storage. The COP of the system is as good as with the cold-water storage and significantly better than with ice storage, as can be expected form the discussions before. The same holds for the running cost, which are only 68 % of the running cost with ice storage, and about the same as with cold-water storage.

9.2.6 Alternative integration concepts

Up to this point, all examples show storages that link the production of cold to the cold demand. However, there are also other options. In the production of cold by any kind of chiller, there is the option of intermediate storage of the waste heat that has to be rejected. In the case of using a heat driven chiller, there is additionally the option to store the driving heat. These options are very significant: they remove the limitation that the phase change temperature of the storage material has to be somewhere between the supply temperature and the demand temperature.

The option of intermediate storage of the waste heat of a chiller is currently investigated by the ZAE Bayern in a demonstration project, supported with funds of the German Federal Ministry of Environment (BMU) under contract number 0329605D. In solar thermal installations with large capacity, full annual utilization is desirable. During the cold season, solar heat serves for space heating. During the warm season, solar heat can be converted into useful cold using a sorption chiller. A favorable situation is when low temperature heating and cooling facilities, e.g. floor or wall heating systems or activated ceilings, are applied for heating and cooling. In that case, a low-temperature heat storage using the latent heat of a PCM can be applied to significantly improve the system in the heating and in the cooling mode. In the heating mode, the heat storage is used for leveling the highly variable solar gain. In the cooling mode, the storage is used in combination with a dry cooling tower in order to replace a wet cooling tower. The typical disadvantages of wet cooling towers, like significant use of water, legionella, fog, and necessary water treatment are thereby eliminated. In combination with the dry cooling tower, the size of the dry cooling tower is reduced.

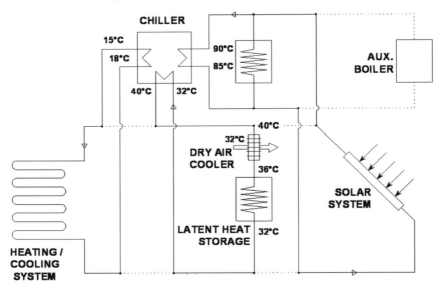

Fig. 9.44. System layout for cooling in summer during daytime operation: waste heat is rejected to the storage and the dry air cooler (picture: ZAE Bayern).

Fig.9.44 shows the system layout for cooling in summer during daytime operation. Heat from a solar system drives the chiller and the produced cold is delivered to the cooling system. Waste heat is rejected to the storage and the dry air cooler, about 50 % each. The storage is hereby charged at 36 °C inlet and 32 °C outlet temperature. For discharge of the stored waste heat at night the storage and the dry air cooler are connected directly; discharging of the storage is at 22 °C inlet and 25 °C outlet temperature. To allow the storage of waste heat in daytime and the discharge during nighttime, the heat has to be stored in a very narrow temperature range between the charging and discharging temperatures. Therefore, a latent heat storage based on $CaCl_2 \cdot 6H_2O$ with a melting temperature of about 29 °C was developed. The storage, shown in fig.9.45, consists of two storage containers made of HDPE with heat exchanger made from capillary tubes with distribution pipes, similar to the storage design of Calmac described in section 9.2.5.1. The storage is filled with about 2.4 t of $CaCl_2 \cdot 6H_2O$ and has a storage capacity of 120 kWh.

Fig. 9.45. Left: heat exchanger made from capillary tubes with distribution pipes. Right: storage containers made of HDPE with the heat exchanger inserted (pictures: ZAE Bayern).

Stand-alone tests on the storage confirmed that the storage achieves the design capacity of 120 kWh over repeated cycles. The charging power was about 12 kW as designed, however the discharging power was somewhat lower than desired. Since summer 2007, the storage is integrated into the system for solar heating and cooling at the ZAE Bayern and the system performance is monitored. This includes also the operation in winter, when the storage is used to store solar heat for heating. First monitoring data recorded since summer 2007 confirm the general feasibility and the thermal design of the latent heat storage and the overall system. The first descriptions of the system have been published by Mehling et al. 2006 and Keil et al. 2007, including system layout and test results from the storage in a stand-alone test. Helm et al. 2007 have presented first test results within the heating and cooling system, and cost calculations of the storage.

This example shows the importance of understanding a whole system to optimize its performance. Often the solution is not in developing a better storage or storage material, but in the optimization of an existing storage within an improved integration concept.

9.3 Examples for space heating

Historically, the application of PCM for space heating has been investigated earlier than for space cooling. However, the experience of the last decade has shown that applications for space cooling were on the market first. This even holds if cold storage using water / ice is not counted. To understand this, it is necessary to discuss why a latent heat storage should be used in an application for space heating or cooling. This comprises actually two questions, the question is if a heat storage should be used, and the question if it should be a latent heat storage or a sensible heat storage, for example using water as storage material. As a boundary condition, it is necessary to take into account that, with the exception of developing countries, almost all buildings are connected to the energy supply grid in some way such that electricity, natural gas, or oil are available. Using these energy forms, it is possible with little equipment to supply heat at a comparatively high power. To supply cold at sufficient power usually causes higher cost for the necessary equipment. This explains why the installation of an ice storage in an air-conditioning system can significantly lower the investment cost. The first question, if a heat storage should be used in an application or not, can therefore be answered positively for cooling often just by looking at the investment cost. The second question is if a storage should be a latent heat storage or a sensible heat storage. In space cooling applications the temperature difference useful for storage is a few to maybe 15 °C only. This favors latent cold storages as the discussed examples show. In heating applications, especially when fossil fuels are burned, the heat can be supplied at high temperatures and the temperature difference between supply and demand often allows already significant storage densities with hot water heat storages (section 9.1.4). This does not hold for solar heating: the higher the temperature of the heat transfer fluid supplied by the solar collector to the storage, the lower the collector efficiency. Further on, in solar heating the use of PCM offers additional design options: the absorber for solar radiation and the storage unit can be combined in a way that is not possible when using sensible heat storage. This is shown by the next examples.

9.3.1 Solar wall

Wall elements for exterior walls to heat a building have been investigated for several decades. On an ordinary wall, the temperature gradient within the wall results from heat loss from the heated interior to the cold outside. Solar heat input by absorption of solar radiation at the surface of the wall is lost easily through free convection or forced convection in case of wind (fig.9.46, left).

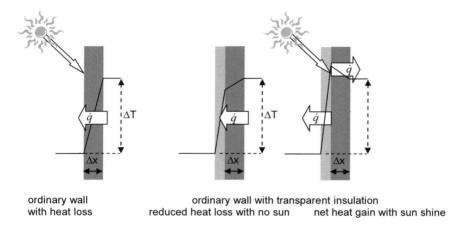

ordinary wall
with heat loss

ordinary wall with transparent insulation
reduced heat loss with no sun   net heat gain with sun shine

Fig. 9.46. Ordinary wall, and wall construction with transparent insulation to use solar gains. If the solar gains are high enough, a net heat input into the building results.

The loss of heat can be reduced using a transparent insulation (fig.9.46, centre). Additionally, the transparent insulation transmits solar radiation, which is then absorbed at the surface of the wall. Because the transparent insulation also insulates, the loss of the heat to the outside is reduced and can even lead to a net heat gain to the building interior (fig.9.46, right). In that case, the wall on the inside of the building acts as a low temperature heater. To be transparent and at the same time an insulation, a transparent insulation consists of a transparent material with a low thermal conductivity (transparent insulation material = TIM) or a layered structure with a high heat transfer resistance. Transparent insulations are state of the art and available as commercial products. The combination of transparent insulation with ordinary walls however has a disadvantage: to ensure enough storage capacity, the wall needs to have a sufficient thermal mass resulting in a wall of significant thickness. This is where PCM offer a unique solution: a thin layer of PCM can replace a thick massive wall. Within the R&D project "Innovative PCM technology" (funded by the German Ministry of Economics and Labor BMWA, contract number 0327303) the Glaswerke Arnold and the ZAE Bayern investigated this approach: the ordinary wall in the construction was replaced by a PCM-wall. Fig.9.47 shows a sketch of the solar wall on the left and the constructed PCM-wall using macroencapsulated PCM on the right.

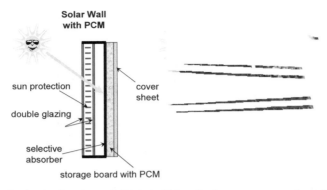

Fig. 9.47. Left: sketch of a solar wall. Right: PCM-wall using macroencapsulated PCM (pictures: ZAE Bayern).

The PCM-wall covered by a double-glazing as transparent insulation was installed in a test building and monitored for several moths. Fig.9.48 shows experimental data recorded during a few days in February.

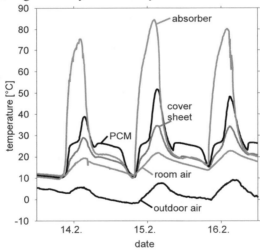

Fig. 9.48. Experimental results of the solar wall with PCM from an installation in a test building (picture: ZAE Bayern).

The experimental data show that as soon as the absorber is heated above about 25 °C the PCM starts to absorb heat and melt. The melting is completed before the absorber temperature falls below the melting temperature again. After a short time, where the PCM is subcooled, it releases the heat slowly throughout the night at a temperature well above 20 °C. That means, if the room air were to be held at 20 °C, the wall would supply heat to the room most of the time. In the experiment, the air in the room cooled down well below 20 °C and consequently the temperature of the cover sheet on the inner side of the wall dropped below 20 °C in the early morning hours.

9.3.2 Daylighting element

If the PCM is partly transparent, the whole element can also be constructed in a way that it transmits light and illuminates the building interior. This idea has been investigated within the R&D project "Innovative PCM technology" by the company Glaswerke Arnold and the ZAE Bayern.

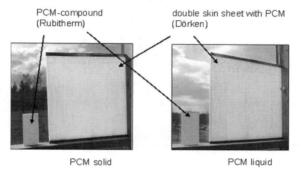

Fig. 9.49. Concepts for a transparent PCM sheet for the integration in a daylighting element (pictures: ZAE Bayern).

Fig.9.49 shows two concepts investigated as transparent PCM sheet. Both, the compound from Rubitherm and the double skin sheet from Dörken show a transparency which changes to some degree between the solid and liquid state of the PCM, and which can be used for a diffuse illumination of the room behind the wall element. Fig.9.50 shows a sketch of the complete system.

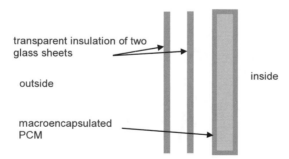

Fig. 9.50. Sketch of the complete system.

The system has been tested at the south façade of an external test room at the ZAE Bayern. Fig.9.51 shows the temperatures measured inside the room, outside, and at the PCM.

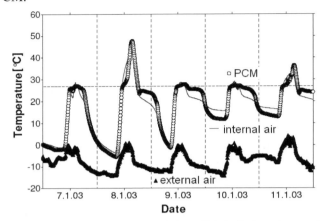

Fig. 9.51. Experimental results from a test of the PCM-dayligting element (picture: ZAE Bayern).

Until 9th of January, the external test room was not heated at all, causing internal temperature lows, which are not typical. The following days, an internal heater kept the room temperature within a normal range and the PCM temperature varied closer around the phase change temperature of 27 °C. Fig.9.52 shows test results for the same days for the globally incident radiation and the radiation transmitted by the PCM-dayligting element. The results show that a large fraction of the radiation is transmitted into the room behind the element for illumination.

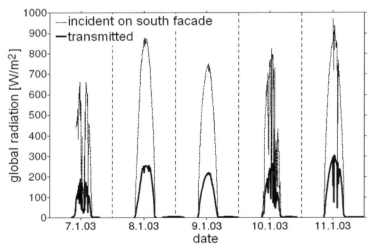

Fig. 9.52. Test results of the globally incident radiation and the radiation transmitted by the PCM-dayligting element (picture: ZAE Bayern).

Detailed information on the development and the testing of the system is available in the publication by Weinläder 2003. This also includes the characterization of the optical properties of PCM, which is a crucial point for this system.

Based on own ideas, the Swiss architect Dietrich Schwarz commercialized a system with his own company GLASSX AG, using PCM from Dörken. The wall element, called GLASSX®crystal, consists of four functional units: a transparent insulation, a protection from overheating, an absorber, and the heat storage. The transparent insulation consists of three glass sheets with a total U-value of 0.5 W/m²K. To protect the element from overheating, a prismatic layer reflects light from the sun in summer incident at high angles, and transmits sunlight in winter when it comes at low angles. The PCM used is the salt hydrate DELTA®-COOL 28 from Dörken with a melting range of 26 – 28 °C and a storage capacity of 75 Wh/liter. The PCM is macroencapsulated in special plastic containers. The whole wall element has only a thickness of 78 mm and is able to store 1185 Wh/m². Using solar energy, the system can heat with inside surface temperatures in the 26 – 28 °C range, and at the same time gives homogenous and diffuse illumination.

Fig. 9.53. GLASSX®crystal installed in a building, on the left as seen from the inside, and on the right as seen from the outside (picture: GLASSX AG).

Fig.9.53 shows the appearance of the system from the inside and from the outside of a building. More information on the GLASSX®crystal wall element, and on existing installations, can be found at the website of the company GLASSX AG.

A similar system has also been investigated by the company Inglas. More information can be found at http://www.inglas.de/glas/heiz/pcm.html.

9.3.3 *Floor heating systems*

Like wall heating systems, floor heating systems have the advantage of a large heat transfer area. Additionally, the heat transfer coefficient for heating from the floor by free convection is even higher. This allows a high energetic efficiency, as the necessary heating power is achieved at a lower surface temperature. The combination of the heating by warm air with thermal radiation also improves the comfort feeling. Besides heating with hot air, heating with electricity and with hot water, especially from solar collectors, is possible.

9.3.3.1 Floor heating system with hot water

A floor heating system using hot water as heat transfer fluid and PCM for heat storage has been developed by the company Rubitherm Technologies GmbH. Fig.9.54 shows the layered structure of the system. The heating pipes are embedded in the storage material, a granulate filled with a paraffin which was described in section 2.4.2.1.

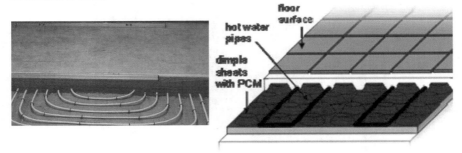

Fig. 9.54. Floor heating systems using hot water for heat supply. Left, using a granulate (picture: Rubitherm Technologies GmbH) and right, using a dimple sheet.

Even though the floor thickness is reduced, 0.5 kWh/m^2 of heat can be stored in the floor. An additional advantage of the concept is that no drying of any component after installation is necessary, thereby reducing installation time and cost compared to common systems.

The company TEAP has developed a similar system, shown at the right in fig.9.54. Here a dimple sheet with a salt hydrate (fig.2.28) is used to integrate the PCM into the floor.

9.3.3.2 Floor heating system with electrical heating

A floor heating system with electrical heating has been developed by the company Sumika Plastech / Japan and is described on their webpage http://www.sumikapla.co.jp/e/main_pd05.html. The working principle is similar to the example above. Instead of water pipes embedded in PCM-granulate, their design consists of electrical heating wires in contact with especially designed PCM-modules. The modules are flat plastic containers filled with a salt hydrate.

Recently there have also been investigations to use shape-stabilized PCM in floor heating systems. Kunping et al. 2005 and Yinping et al. 2006 describe an electric floor heating system using shape-stabilized paraffin as storage medium. Fig.9.55 shows the layered structure consisting of the electrical heater, the plates of shape-stabilized paraffin, and a wooden floor cover.

Fig. 9.55. Layered structure of an electric floor heating system. From left to right: electric heaters, wooden frame with plates of shape-stabilized PCM, and wooden floor cover (pictures: Tsinghua University, PR China)

An experimental building was equipped with the system and tests were performed. One of the key findings was that the system allowed more than half of the electricity demand to be shifted from the on-peak period to the off-peak period. This can be a significant economic benefit when using off-peak electricity at a lower price.

9.3.3.3 Floor heating system using hot air

Floor heating systems using hot air have already been in use in Roman times. The Romans used hot air from wood fires and circulated the hot air through channels under the floor surface. The ceramic tiles used for the floor construction were able to store some heat and thereby assured comfortable room temperatures even after the fire was out.

Within the R&D project "Innovative PCM technology" the company Grammer and the ZAE Bayern investigated the application of PCM as storage material in the floor structure heated by hot air supplied by solar air collectors. Fig.9.56 shows a sketch of a solar air collector.

Fig. 9.56. Sketch of a solar air collector (picture: Grammer).

The investigation of several approaches to integrate the PCM however did not result in a suitable solution regarding storage capacity and heat transfer under the boundary conditions given for floor heating.

9.3.4 Solar air heating and ventilation system

All heating systems discussed up to this point heat a room by heating one of the room surfaces and the heat is supplied electrically, by hot water, or by hot air. For the case of heating with hot air from solar air collectors, there is also a different option. The heating can be combined with the supply of fresh air in an energy-efficient manner: fresh air is supplied to the solar air collector, heated, and finally supplied to the building as fresh and warm air. A problem using air as heat transfer medium is that air stores only little heat due to its low heat capacity and temporal differences between heat demand and available solar heat become a problem. The use of PCM seems to be very promising because the heat storage density is high, and because PCM will further on have a regulating effect on the supply air temperature. Within the R&D project "Innovative PCM technology" the company Grammer and the ZAE Bayern investigated the application of PCM in such a solar air system. Besides the integration into a floor-heating unit discussed above, the integration of PCM was also investigated in the collector and absorber unit, the ventilation pipes, and as separate storage connected in the air ducts between the collector and the building interior. For the last option, the separate storage, a prototype was built and tested in a real environment. The prototype, shown in fig.9.57, is a module type storage.

Fig. 9.57. Prototype of the developed storage (picture: Grammer).

The storage is installed in a solar air system at a building owned by the company Grammer. It is charged in daytime by 20 m² collectors and discharged at nighttime. The storage was continuously operated from February 2003 until December 2007. Since February 2004, 19 ClimSel™ Thermal Batteries, shown at the left in fig.2.30, were used as storage modules. They are filled with C32, a salt hydrate with a melting temperature of 32 °C. Temperatures, airflow, and the stored and discharged heat were recorded. Only little subcooling, and no loss in the performance of the storage modules was observed. Upon storage of heat supplied by the collector, the temperature peaks were buffered by the storage. Upon discharge of heat, the temperature of the air supplied to the building was raised by the storage by about 5 to 8 K.

A similar system was designed and tested as a full-scale prototype at the Sustainable Energy Centre of the University of South Australia. Bruno and Saman 2002 present the results of charging and discharging tests on the storage and compare them to those from a computer model. The storage uses flat sheets of encapsulated PCM with a melting temperature of 29 °C. The total amount of PCM was 253 kg. In the discharging test, that is when the storage is used for heating, a discharging power of more than 1.75 kW was observed over the first 6 hours. During this time, the temperature of the discharged air was above 23 °C and therefore sufficient to fulfill the comfort demand at all times. It is planned to operate the system for several years.

Another similar system has been developed at the Faculty of Mechanical Engineering, University of Ljubljana. Arkar and Medved 2002 present the design and experimental testing of a model heat storage and calculations to design a real scale system. In the system design, the storage is again integrated into the duct of the building ventilation system. Ambient air is first heated in the solar roof, then passes trough latent heat storage, and finally enters the living space as preheated ventilation air. The function of the heat storage is to prevent overheating of the air supplied to the building interior in daytime when the solar heat input is high, and to allow ventilation and heating to continue in times with lower solar heat input. To design a real scale system, it was necessary to verify the numerical model used for the storage. For this purpose, a model storage was designed and tested. The storage is a module type storage with polyethylene spheres as encapsulation of the

PCM (fig.5.19). Approximately 90 % of the sphere volume is filled with PCM and the spheres occupy 62 % of the storage volume, giving a total PCM fraction of 56 vol.%. After verification of the numerical code with experimental results, calculations to design a real scale system were performed. The results show that with a well designed storage it is possible to prevent or to reduce overheating during the day and that the system can supply fresh warm air to the building interior even until the next morning.

9.3.5 Storage for heating with hot water

Many heating systems installed today use hot water as heat transfer medium. In heating systems with ordinary heating units inlet temperatures are usually around 60 °C and return temperatures around 40 °C. This is independent whether the heat source is solar or a heater burning fossil fuel. Conventional heating systems use hot water heat storages. They can have different designs: with heat exchanger or with direct discharge, and with or without stratification. In order to save space by reducing the storage size or to be able to store more heat in a given volume, for example to extend the time where a solar heating system can operate without sun, storages concepts which offer higher storage density are necessary. The main approach to get higher storage density is to use latent instead of sensible heat storage. Due to the good heat transfer coefficient to water of several 100 W/m^{2}K and the high heat capacity of water, the bottleneck for heat transfer is usually within the PCM. This has to be taken into account in the storage design.

9.3.5.1 Heat exchanger type approach

The most common approach to design a latent heat storage for space heating is using a heat exchanger type, usually with pipes. To improve the heat transfer in the storage medium, fins can be attached to the heat exchanger. For space heating, such a system is commercially available from the company Alfred Schneider GmbH. The storage LWS 750, which is shown on the left in fig.9.58, uses a salt hydrate and is designed to be used with solar collectors, wood heating systems, and other heat sources.

Fig. 9.58. Left: latent heat storage LWS 750 (picture: Alfred Schneider GmbH). Right: model of a heat storage using a PCM-graphite compound (picture: SGL Technologies).

The technical data of the storage are available at http://www.alfredschneider.de/prod05.htm. The storage has a size of 750 x 750 x 1590 mm^3, and a weight of 1165 kg. Its storage capacity is 122 kWh (= 439 MJ) between 100 °C and 40 °C, and its heating power is 1.7 kW at 47 °C inlet and 52 °C return temperature. In the same temperature range, a hot water heat storage can store 246 MJ/m^3. Taking the whole volume of the storage of 0.894 m^3, the hot water heat store can store 220 MJ. In reality, this value is not achieved, because due to the insulation the volume available for the water is significantly smaller and because hot water heat storages are usually not heated to 100 °C because of the vapor pressure. Taking this into account about a 3-times higher storage density can be assumed using LWS 750.

Another concept to improve the heat transfer is using a PCM-graphite-compound as heat storage medium. This compound has been described in section 2.4.2.2. Within the R&D project "Innovative PCM technology" the companies Bosch, SGL Technologies, Behr, Merck, and the ZAE Bayern developed and tested a storage concept using a pipe heat exchanger and the PCM-graphite-compound as heat storage medium. The right picture in fig.9.58 shows a demonstration model in an acrylic glass tank. The PCM used in the PCM-graphite compound is NaOAc·3H$_2$O with a melting temperature of 58 °C. Within the temperature range from 40 °C to 60 °C that is used in conventional heating systems, a hot water heat storage would store about 80 MJ/m^3. The developed latent heat storage has about 250 MJ/m^3, where it is taken into account that due to the graphite and the pipes the PCM only uses about 80 vol.% of the storage space. The storage density in the developed latent heat storage is also about three times as high as for the hot water heat storage. Two prototype storage modules have been tested for more that 1300 cycles without loss of performance.

9.3.5.2 Module type approach

Because of its flexibility in the design, the approach to use a module type storage is also attractive.

Hiroshi 1998 describes a space heating system installed in the North of Japan. The system was installed by Mitsubishi Chemical Engineering Co., who has a license from Cristopia for the Japanese market. The storage is based on the Cristopia design and has a tank volume of 23 m^3 and a storage capacity of 3900 MJ. The PCM used consists of NaOAc·3H$_2$O with additives, and has a melting temperature of 47 °C.

Velraj and Nallusamy 2006 describe investigations on a module type storage integrated in a solar water heating system. The system is composed of the heat storage, a solar flat plate collector, flow meter, and circulating pump. The heat storage consists of an insulated cylindrical tank, which contains the PCM encapsulated in spherical capsules with an outer diameter of 55 mm. The PCM was a paraffin with a melting temperature of 60 °C and latent heat of 213 kJ/kg. The PCM capsules occupy 51 vol.% of the total volume of the tank, the remaining volume is occupied by water. As was mentioned in tab.5.1, module type heat storages are often called hybrid storages when the heat transfer medium is a liquid. The reason is that the significant volume fraction of heat transfer fluid in the storage tank leads to a significant fraction of heat stored as sensible heat. The combination of sensible and latent heat storage has the potential to solve difficulties experienced in systems using sensible heat storage or latent heat storage only. The objective of the investigations was to predict the thermal behavior of such a hybrid storage integrated in a solar water heating system. Therefore, discharge experiments were carried out in two ways: as continuous process, and as batchwise process with a time interval of 10 minutes between batches allowing transfer of heat from the PCM to the water in the tank. Further on, parametric studies to examine the effects of the volume fraction of the PCM and flow rate of the heat transfer fluid were done. The performance of the system was finally compared to a conventional sensible heat storage system. The authors concluded that the combined sensible and latent heat storage concept reduces the size of the storage tank appreciably compared to a sensible heat storage and that it is best suited for applications where the requirement is intermittent, that is for batchwise discharge.

A modified concept of a hybrid storage, especially designed for hot water heat storages with stratification, has been presented by Mehling et al. 2001 and Mehling et al. 2002a. Stratification means that water of different temperatures is separated by gravitation due to the temperature dependent density of water: hot water with the lower density rises to the top of the storage and cold water with the higher density sinks to the bottom. Mixing of hot and cold water can reduce the maximum temperature in the tank below the lowest useful temperature. The advantage of stratification is that hot and cold water are separated within the same tank and not mixed. This is especially an advantage in solar heating systems, because solar collectors supply heat at different temperatures, depending on the solar heat input. Hot water heat storages are often used for space heating and for domestic hot water. Then, the temperatures in a storage can cover the range from 10 °C, caused by incoming domestic water from a pipe in the ground, to 60 °C, necessary for space heating and domestic hot water, or even higher to load the storage with as much heat as possible. While the temperature differences at the storage bottom are large between the loaded and the unloaded state, they are comparatively small at the top for most of the time. This means the integration of PCM in the top can significantly increase the storage density there, whereas at the bottom the water already has a quite high storage density due to the large temperature changes. First experiments on a functional model and a numerical study were published in Mehling et al. 2001 and Mehling et al. 2003. Later, the concept was tested using a modified Hybrid 550 hot water heat storage (fig.9.59), donated by the company Fa. Sailer-Solarsysteme.

Fig. 9.59. Outside view of the test storage with temperature sensors, PCM module, and inside view with PCM modules fixed at the top of the storage (pictures: ZAE Bayern).

The storage with a volume of 550 liters was equipped with PCM modules, which had a total volume of 33 liters and were filled with a $NaOAc \cdot 3H_2O$-graphite compound. In the temperature range from 70 °C to 40 °C the 6 vol.% PCM increase the amount of heat stored by 14 % with reference to the whole storage volume, and by 41 % with reference to the layer defined by the vertical extension of the PCM modules (fig.9.60). One effect of this increase of storage density is that when there is no solar heat input for many days, the storage can deliver hot water from the top section, defined by the vertical extension of the PCM modules, for a period extended by two days.

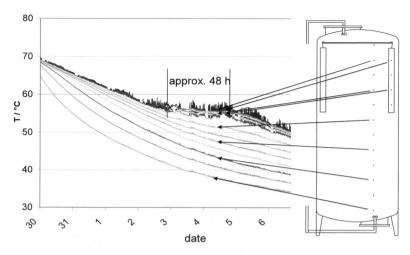

Fig. 9.60. Temperature history at different positions in the water and the PCM in an experiment over several days with heat loss to the ambient (picture: ZAE Bayern).

Further on, discharge experiments were carried out as continuous process and as a batchwise process. In both cases, the PCM could show an improved performance compared to only sensible heat storage. Cabeza et al. 2003 describe investigations performed under real operating conditions in a complete solar heating system that has been constructed at the University of Lleida, Spain. The results show that only 3 kg of the PCM compound in a 146 liter storage tank are enough to compensate for a heat loss of 3-4 °C in 32 liters of water in the top of the storage. This is equivalent to the cool down in the storage without PCM module over a period of about 10 hours. Reheating of cold water after a fast discharge of the storage was in 10-15 minutes. This is considered fast enough for most applications. Recently, a comparison of the stratification between a storage tank with and without PCM-module was done. Both tanks presented a very similar stratification for charging as well as for discharging (Castell et al. 2007).

9.3.5.3 Direct contact type approach

Direct contact type systems using salt hydrates with phase change temperatures suitable for space heating have been investigated already more than two decades ago. As discussed in section 5.4.2.4, such systems are described in detail by Lindner 1984, Fouda et al. 1984, Farid and Yacoub 1989, Farid et al. 2004. However, it seems as if there are currently no activities in this field.

The Galisol-system, also useful for salt hydrates and discussed in section 5.4.2.4, was developed close to market, but never finally made it. The reason is that only CFCs, which are today banned from the market due to environmental reasons, were found to be suitable as heat transfer fluids. A direct contact system using a paraffin as PCM and water as heat transfer fluid has also been developed, but was never commercialized.

9.3.5.4 Slurry type approach

The use of PCM slurries in space heating has also been suggested. Heinz and Streicher 2005, from the Institute of Thermal Engineering at Graz University of Technology, performed experimental tests on a storage tank filled with a microencapsulated PCM slurry. The PCM slurry was supplied by BASF and had a melting temperature of about 60 °C. Slurries of different concentrations were tested as heat storage fluid in a tank with a volume of 200 liters. The storage tank was charged and discharged with different types of heat exchangers and the heat transfer coefficients both for natural (internal heat exchanger) and forced convection (external flat plate heat exchanger) were analyzed. It seems that there is no full-scale installation of a slurry system for space heating so far.

9.4 Further information

Even though the chapter on applications for heating and cooling in buildings is rather comprehensive, there is still some interesting information left that did not fit into the structure of the chapter. This information has been collected at this point.

First of all, an interesting numerical and experimental study of walls and roofs filled with PCM in different layers has been published by Ismail and Castro 1997. Even though usually different locations to apply the PCM are preferred, the paper is a very good introduction to understand the consequences of integrating PCM on the temperature variations at different times and locations.

Regarding systems for free cooling in general, systematic studies on the basics like PCM selection and heat transfer have been done by Zalba et al. 2003, Zalba et al. 2004, and Cabeza et al. 2003b.

Regarding the use of ice storages for space cooling, ASHRAE 1991 can serve as a good introduction. It discusses different system designs and operation modes, economic considerations, and different storage types like cold (chilled) water storage, ice on coil systems, systems with ice in modules, ice harvesting systems, and ice slurry systems.

Many of the new developments for space heating and cooling have been tested in the laboratory, usually under controlled conditions simulating real conditions. However, the simulation of real conditions can be based on wrong assumptions and also does not include the influence of users etc. This is usually problematic when trying to market a new development. For this reason, many companies have demonstration buildings with regular use, where the new developments are tested and monitored. Several examples for single products have been mentioned in the text. To test the performance of the blinds with PCM as well as other systems under realistic conditions, the ZAE Bayern and a number of industry partners currently perform a demonstration project named "Development and practical performance testing of building components with PCM in demonstration buildings", short PCM Demo. The objective of this project is to support the introduction of building products and systems containing PCMs to the market. Active, as well as passive approaches for cooling a building with PCM are tested in administrative and public buildings under realistic use. Besides product development, this research project primarily focuses on presenting buildings with demonstration character. The idea is to support the market launch of PCM systems by making correspondingly equipped reference buildings accessible to the public and presenting them favorably in the media (internet, specialized journals, conferences, etc.). When the system performance has been recorded, simulation tools can be validated and design guidelines compiled. Feedback from building users will provide a direct indication of how the products are accepted, which in turn will point out valuable optimization potential to manufacturers and planners. Evaluating the metrological data with respect to saving energy will enable energy suppliers to incorporate PCM systems into their energy consultations. Energy suppliers will also be in a position to evaluate the effects of wide-scale use of PCM systems on power plant and grid capacities by carrying out an economic analysis based on the load shift measured. The project, started in 2006, continues to the end of 2009. Up to date information is available at http://www.pcm-demo.info/objective.htm. Furtheron, this project is part of a larger project called LowEx (www.lowex.info), which focuses on heating and cooling with low exergy in general, funded by the German Ministry of Economics (BMWi).

A product not mentioned before, but actually dealing with space cooling in a wider sense, is a cold storage for a truck parking air-conditioning system. Air-conditioning during driving breaks is often necessary to allow drivers to relax or sleep. It is very common to let the engine running in order to have sufficient electricity for the air-conditioning system. This is economically and environmentally not a good solution. To solve this problem, the company Webasto AG has developed a cold storage. The cold storage is loaded with the compression cooler when the truck is driving, and when the truck is parked only the storage is used for air-conditioning. In this case, the cold is produced in daytime and used at nighttime, in contrast to the applications discussed before. The increase of the efficiency is not realized by the efficiency of the chiller, but by the engine producing the electricity that is not running in stand by anymore. The developed cold storage uses water as phase change material and a graphite matrix to improve the heat transfer. This approach makes it possible to have cooling powers of up to 1 kW just from the cold storage. Compared to earlier versions of cold storages this is a four times higher power per volume. More information on the system design and test results was presented at the 6th Karlsruher Fahrzeugklima-Symposium "Standklimatisierung und Wärmepumpenheizung im KFZ" and can be downloaded at http://www.twk-karlsruhe.de/DownloadSymposium170904.html. The cold storage is produced since late 2005 and is at the time of writing this book promoted for the American market as Webasto BlueCool Truck.

9.5 References

[Alfred Schneider GmbH] Alfred Schneider GmbH. Germany. http://www.alfredschneider.de
[Arkar and Medved 2002] Arkar C., Medved S.: Enhanced solar assisted building ventilation system using sphere encapsulated PCM thermal heat storage. Presented at 2nd Workshop IEA ECES Annex 17 "Advanced thermal energy storage techniques – Feasibility studies and demonstration projects", Ljubljana, Slovenia, 3-5 April 2002. www.fskab.com/annex17
[ASHRAE 1991] 1991 ASHRAE Handbook, HVAC Applications, SI Edition, ASHRAE Atlanta, 1991, ISBN 0-910110-79-4
[Axima] Axima Refrigeration GmbH. Germany. www.axima-ref.de
[BASF AG] BASF AG. Ludwigshaven, Germany. www.micronal.de
[Braun 2003] Braun J.E.: Load control using building thermal mass. Journal of Solar Energy Engineering **125**(3), 292-301 (2003)
[Bruno and Saman 2002] Bruno F., Saman W.: Testing of a PCM energy storage system for space heating. Proc. of World Renewable Energy Congress VII, Cologne, Germany, 2002
[BTC Speciality Chemical Distribution GmbH] BTC Speciality Chemical Distribution GmbH. Cologne, Germany. www.btc-europe.com
[BUCO Wärmeaustauscher GmbH] BUCO Wärmeaustauscher GmbH. Geesthacht, Germany. www.buco-gmbh.de
[Cabeza et al. 2003a] Cabeza L.F., Noques M., Roca J., Illa J., Hiebler S., Mehling H.: PCM-module to improve hot water heat stores with stratification: first tests in a complete solar system. Proc. of Futurestock, 9th International Conference on Thermal Energy Storage, Warsaw Poland, Sept 2003
[Cabeza et al. 2003b] Cabeza L.F., Zalba B., Marin J.M., Mehling H.: PCM-graphite matrix in flat plate encapsulates for low temperature applications. Proc. of Futurestock, 9th International Conference on Thermal Energy Storage, Warsaw Poland, Sept 2003
[Cabeza et al. 2007] Cabeza L.F., Castellón C., Nogués M., Medrano M., Leppers R., Zubillaga O.: Use of microencapsulated PCM in concrete walls for energy savings. Energy and Buildings **39**, 113–119 (2007)
[CALMAC Manufacturing Corp.] CALMAC Manufacturing Corp. Fair Lawn, USA. www.calmac.com
[Castell et al. 2007] Castell A., Solé C., Medrano M., Nogués M., Cabeza L.F.: Comparison of stratification in a water tank and a PCM-water tank. Proc. of Energy Sustainability 2007, Long Beach, California, June 27-30, 2007
[Çengel 1998] Çengel Y.: Heat transfer – A practical approach. Mc Graw Hill (1998)
[Climator AB] Climator AB. Sweden. http://www.climator.com
[Cristopia] Cristopia Energy systems. Vence, France. www.cristopia.com
[Darbouret et al. 2005] Darbouret M., Cournil M., Herri J.-M.: Crystallisation and Rheology of an hydrate slurry as secondary two-phase refrigerant for air-conditioning application. Proc. of 6th Workshop on Ice Slurries of the International Institute of Refrigeration, Yverdon-les-Bains, Switzerland, 15 - 17 June 2005
[Detzer and Boiting 2004] Detzer R., Boiting B.: PCM eröffnet neue Wege für die Raumlufttechnik. KI Luft- und Kältetechnik **9**, 350-352 (2004)
[Dincer and Rosen 2002] Dincer I., Rosen M.A.: Thermal energy storage. Systems and applications. John Wiley & Sons, Chichester (2002)

[Dörken] Dörken GmbH & Co.KG. Herdecke, Germany. www.doerken.de
[DuPont] DuPont. Luxembourg. www.energain.dupont.com
[Egolf et al. 2003] Egolf W., Sari O., Vuarnoz D., Caesar D.A., Sletta J.: A review from physical properties of ice slurry to industrial applications. Proc. of Phase change material & slurry engineering conference & business forum, Yverdon-les-Bains, Switzerland, 2003
[Emco] Emco. Lingen, Germany. www.emco-klima.de
[Farid and Yacoub 1989] Farid M.M., Yacoub K.: Performance of direct contact latent heat storage unit. Solar energy 43, 237-252 (1989)
[Farid et al. 2004] Farid M.M., Khudhair A.M., Razack S.A.K., Al-Hallaj S.: A review on phase change energy storage: materials and applications. Energy Conversion and Management 45, 1597–1615 (2004)
[Fouda et al. 1984] Fouda A.E., Despault G.J., Taylor J.B., Capes C.E.: Solar storage system using salt hydrate latent heat and direct contact heat exchange – II, characteristics of pilot operating with sodium sulphate solution. Solar energy 32, 57-65 (1984)
[GlassX AG] GlassX AG. Zürich, Switzerland. www.glassx.ch
[Gschwander and Schossig 2006] Gschwander S., Schossig P.: Phase Change Slurries as heat storage material for cooling applications. Proc. of ECOSTOCK, 10th International Conference on Thermal Energy Storage, Stockton USA, 2006
[Haussmann and Schossig 2006a] Haussmann T., Schossig P.: Baustoffe mit Phasenwechselmaterialien als Kältespeicher für energieeffiziente Gebäude: Statusseminar "Thermische Energiespeicherung – mehr Energieeffizienz zum Heizen und Kühlen", Freiburg, 2.-3. November 2006
[Haussmann and Schossig 2006b] Haussmann T., Schossig P.: PCM-Aktiv: A project for active driven construction materials with latent heat storage. Proc. of 7th Conference on Phase Change Materials and Slurries for Refrigeration and Air Conditioning, Dinan, France, 13 - 15 September 2006
[Hawes et al. 1993] Hawes D.W., Feldmann D., Banu D.: Latent heat storage in building materials. Energy and buildings 20, 77-86 (1993)
[He et al. 1999] He B., Gustafsson E.M., Setterwall F.: Tetradecane and hexadecane binary mixtures as phase change materials (PCMs) for cool storage in district cooling system. Energy 24, 1015-1028 (1999)
[He and Setterwall 2002] He B., Setterwall F.: Technical grade paraffin waxes as phase change materials for cool thermal storage and cool storage systems capital cost estimation. Energy Conversion and Management 43, 1709–1723 (2002)
[Heinz and Streicher 2005] Heinz A., Steicher W.: Experimental testing of a storage tank filled with microencapsulated PCM slurries. Proc. of 2nd Conference on Phase Change Material & Slurry : Scientific Conference & Business Forum, Yverdon-les-Bains, Switzerland, 15 – 17 June 2005
[Helm et al. 2007] Helm C. Keil, M. Helm, S. Demel, H. Köbel, S. Hiebler, H. Mehling, C. Schweigler: Solares Heizen und Kühlen mit Absorptionskältemaschine und Latentwärmespeicher – Erfahrungen des ersten Betriebsjahres. Proc. of Deutsche Kälte- und Klimatagung 2007, Hannover, 21.-23.11.2007.
[Hiroshi 1998] Hiroshi I.: Operation Records of Encapsuled Heat Storage System. Presented at 2th Workshop IEA ECES Annex 10 "Phase change materials and chemical reactions for thermal energy storage", Sofia, Bulgaria, 11–13 April 1998. www.fskab.com/annex10
[ILKAZELL] ILKAZELL Isoliertechnik GmbH Zwickau. Zwickau, Germany. www.ilkazell.de
[Imtech] Imtech Deutschland GmbH & Co. KG, Hamburg, Germany. www.imtech.de
[Ismail and Castro 1997] Ismail K.A.R., Castro J.N.C.: PCM thermal insulation in buildings. Int. J. Energ. Res. 21, 1281 – 1296 (1997)
[Johansson et al. 2006] Johansson C., Martin V., He B., Setterwall F.: Distributed high capacity cold storage in district cooling systems. Proc. of Ecostock, 10th International Conference on Thermal Energy Storage, Stockton, USA, 2006
[Kalz et. al 2006] Kalz D., Pfafferott J., Schossig P., Herkel S.: Thermally Activated Building Systems Using Phase-Change-Materials. Proc. of EUROSUN 2006

[Keil et al. 2007] Keil C., Helm M., Demel S., Köbel H., Hiebler S., Mehling H., Schweigler C.: Design and Operation of a Solar Heating and Cooling System with Absorption Chiller and Latent Heat Storage. Proc. of OTTI 2nd International Conference on Solar Air-conditioning, Tarragona, Spain Oct. 2007

[Kunping et al. 2005] Kunping Lin., Yinping Zhang., Xu Xu., Hongfa Di., Rui Yang., Penghua Qin: Experimental study of under-floor electric heating system with shape-stabilized PCM plates. Energy and Buildings **37**, 215-220 (2005)

[Lane 1983] Lane G.A.: Solar Heat Storage: Latent Heat Material - Volume I: Background and Scientific Principles. CRC Press, Florida (1983)

[Lindner 1984] Lindner F.: Latentwärmespeicher Teil 1: Physikalisch-technische Grundlagen. Brennst.-Wärme-Kraft, **36**, (1984)

[Lüdemann 2006] Lüdemann B.: Latentwärmespeicher in Lüftungsgeräten: Erste Ergebnisse aus dem Imtech-Haus. Tagungsband: LowEx - Heizen und Kühlen mit Niedrig-Energie, Symposium zum Verbundvorhaben des Bundesministeriums für Wirtschaft und Technologie; Hamburg, Imtech-Haus, 4. Oktober 2006

[Lüdemann and Detzer 2007] Lüdemann B., Detzer R.: Aktive Raumkühlung mit Nachtkälte – Entwicklung eines dezentralen Lüftungsgerätes mit Latentwärmespeicher, KI Kälte – Luft – Klimatechnik, April 2007

[MacCracken 2004] MacCracken M.M.: Thermal Energy Storage and Sustainable Buildings. ASHRAE Journal Sept 04. www.calmac.com

[MacCracken 2006] MacCracken M.M.: The Commercialization of TES in the USA. Proc. of ECOSTOCK, 10th International Conference on Thermal Energy Storage, Stockton, USA, 2006

[maxit] maxit Deutschland GmbH, Germany, http://www.maxit.de

[Mehling 2001] Mehling H.: R&D project "Innovative PCM-Technology". Presented at 8th Workshop IEA ECES Annex 17 "Advanced thermal energy storage techniques – Feasibility studies and demonstration projects", Garching, Germany, Okt. 2001. www.fskab.com/annex17

[Mehling et al. 2001] Mehling H., Cabeza L.F., Hippeli S., Hiebler S.: PCM-module to improve hot water heat stores with stratification. Presented at 8th Workshop IEA ECES Annex 17 "Advanced thermal energy storage techniques – Feasibility studies and demonstration projects", Garching, Germany, Okt. 2001. www.fskab.com/annex17

[Mehling 2002] Mehling H.: News on the application of PCMs for heating and cooling of buildings. Presented at 3rd Workshop IEA ECES Annex 17 "Advanced thermal energy storage techniques – Feasibility studies and demonstration projects", Tokyo, Japan, Okt. 2002. www.fskab.com/annex17

[Mehling et al. 2002a] Mehling H., Cabeza L.F., Hiebler S., Hippeli S.: Improvement of stratified hot water heat stores using a PCM-module. Proc. of EuroSun 2002, Bologna, Italy, June 2002

[Mehling et al. 2002b] Mehling H., Manara J., Körner W.: Potential to improve the thermal comfort in buildings using latent heat storage materials under climatic conditions of Germany. Proc. of WREC 2002, Cologne, Germany, 2002

[Mehling et al. 2003] Mehling H., Cabeza L.F., Hippeli S., Hiebler S.: PCM-Module to improve hot water heat stores with stratification. Renewable energy **28 / 5**, 699 – 711 (2003)

[Mehling et al. 2006] Mehling H., Hiebler S., Schweigler C., Keil C.: Development of a Latent Heat Storage for a Solar Heating and Cooling System. Proc. of Eurosun 2006, Glasgow, Schottland, June 2006

[Mehling et al. 2007] Mehling H., Cabeza L.F., Yamaha M.: Application of PCM for heating and cooling of buildings. In: Paksoy H.Ö. (ed.): Thermal energy storage for sustainable energy consumption – fundamentals, case studies and design, pp. 323-348. Springer, (2007), NATO Science series II. Mathematics, Physics and Chemistry – Vol. 234, ISBN 978–1–4020–5289–7

[Merker et al. 2002] Merker O., Hepp F., Beck J., Fricke, J.: A New Solar Shading System with Phase Change Material (PCM). Proc. of World Renewable Energy Congress VII, Cologne, Germany (2002)

[Nagano 2007] Nagano K.: Development of the PCM floor supply air-conditioning system. In: Paksoy H.Ö. (ed.): Thermal energy storage for sustainable energy consumption – fundamentals, case studies and design, pp. 367-373. Springer, (2007), NATO Science series II. Mathematics, Physics and Chemistry – Vol. 234, ISBN 978–1–4020–5289–7

[Neeper 2000] Neeper D.A.: Thermal dynamics of wallboard with latent heat storage. Solar Energy, Vol. 68, No. 5, 393-403 (2000)

[Nordell and Skogsberg 2007] Nordell B., Skogsberg K.: The Sundsvall snow storage – six years of operation. In: Paksoy H.Ö. (ed.): Thermal energy storage for sustainable energy consumption – fundamentals, case studies and design, pp. 349-366. Springer, (2007), NATO Science series II. Mathematics, Physics and Chemistry – Vol. 234, ISBN 978–1–4020–5289–7

[Pollerberg et al. 2005] Pollerberg C., Noeres P., Dötsch C.: PCS-Systems in cooling and cold supply networks. Proc.of 6th Workshop on Ice Slurries of the International Institute of Refrigeration, 15 - 17 June 2005 Yverdon-les-Bains, Switzerland

[Rubitherm] Rubitherm Technologies GmbH. Berlin. http://www.rubitherm.com

[Sari et al.] Sari O., Egolf P.W., Lugo R., Fournaison L.: Direct contact evaporation applied to the generation of ice slurries to the generation of ice slurries modelling & experimental results modelling & experimental results. http://www.sesec.org/pdf/Dr%20Osmann%20Sari.pdf

[SGL Technologies GmbH] SGL TECHNOLOGIES GmbH. Meitingen, Germany. www.sglcarbon.com/eg

[Shapiro et al. 1987] Shapiro M., Feldman D., Hawes D., Banu D.: PCM thermal storage in drywall using organic phase change material. Passive Solar J **4**, 419-438 (1987)

[Shibutani 2002] Shibutani: PCM-micro Capsule Slurry Thermal Storage System for Cooling in Narita Airport. Presented at 3rd Workshop IEA ECES Annex 17 "Advanced thermal energy storage techniques – Feasibility studies and demonstration projects", Tokyo, Japan, Okt. 2002. www.fskab.com/annex17

[Skogsberg 2005] Skogsberg K.: Seasonal Snow Storage for Space and Process Cooling. Doctoral Thesis 2005:30, Division of Architecture and Infrastructure, Luleå University of Technology; ISSN: 1402-1544. http://epubl.luth.se/1402-1757/2001/51/LTU-LIC-0151-SE.pdf

[Skogsberg and Nordell 2006] Skogsberg K., Nordell B.: Snow cooling in Sweden. Proc. of ECOSTOCK, 10th International Conference on Thermal Energy Storage, Stockton, USA, 2006

[Stetiu 1999] Stetiu C.: Energy and peak power potential of radiant cooling systems in US commercial buildings. Energy and Buildings 30(2), 127-138 (1999)

[Takeda et al. 2003] Takeda S., Nagano K., Mochida T., Nakamura T.: Development of floor supply air conditioning system with granulated phase change materials. Proc. of Futurestock, 9th International Conference on Thermal Energy Storage, Warsaw, Poland, 2003

[Takeda et al. 2005] Takeda S., Nagano K., Nakayama K., Shimakura K., Nakamura T.: Heating operation in the PCM floor supply air conditioning system. Presented at 8th Workshop IEA ECES Annex 17 "Advanced thermal energy storage techniques – Feasibility studies and demonstration projects", Kizkalesi, Turkey, 18-20 April 2005. www.fskab.com/annex17

[Takeda et al. 2006] Takeda S., Nagano K., Nakayama K., Shimakura K.: Study on a floor supply air conditioning system with thermal energy storage using granulated PCM. Proc. of ECOSTOCK, 10th International Conference on Thermal Energy Storage, Stockton, USA, 2006

[Ulvengren 2005] Ulfvengren R.: Passive cooling. Presented at 8th Workshop IEA ECES Annex 17 "Advanced thermal energy storage techniques – Feasibility studies and demonstration projects", Kizkalesi, Turkey, 18-20 April 2005. www.fskab.com/annex17

[Velraj et al. 2002] Velraj R., Anbudurai, K., Nallusamy, N., Cheralathan, M.: PCM based thermal storage system for building air conditioning at Tidel Park, Chennai. Proc of WREC, Cologne, Germany, 2002

[Velraj and Nallusamy 2006] Velraj R., Nallusamy N.: Experimental investigation on a combined sensible and latent heat storage unit integrated with solar water heating system. Proc. of ECOSTOCK, 10th International Conference on Thermal Energy Storage, Stockton, USA, 2006

[Weinläder 2003] Weinläder H.: Optische Charakterisierung von Latentwärmespeichermaterialien zur Tageslichtnutzung. Dissertation at Bayerische Julius-Maximilians-Universität Würzburg, Würzburg, Juli 2003

[Yamaha and Misaki 1998] Yamaha M., Misaki S.: An evaluation of PCM storage installed in air distribution systems - Thermal characteristics of a storage tank with organic hydrate (PCM-13) and a simulation model. Presented at 2th Workshop IEA ECES Annex 10 "Phase change materials and chemical reactions for thermal energy storage", Sofia, Bulgaria, 11–13 April 1998. www.fskab.com/annex10

[Yamaha 2001] Yamaha M.: A study on a heat exchanging ventilation system for residential house using phase change material. Presented at Planning Workshop for IEA ECES Annex 17 "Advanced thermal energy storage techniques – Feasibility studies and demonstration projects", Lleida, Spain, 5-6 April. www.fskab.com/annex17

[Yinping et al. 2006] Yinping Z., Guobinga Z., Ruib Y., Kunpinga L.: Our research on shape-stabilized PCM in energy-efficient buildings; Proc. of ECOSTOCK, 10th International Conference on Thermal Energy Storage, Stockton, USA, 2006

[Zalba et al. 2003] Zalba B., Marin J.M., Sanchez-Valverde B., Cabeza L.F., Mehling H.: Free cooling; an application of PCMs in TES". Presented at 6th Workshop of IEA ECES Annex 14, Lleida, Spain, 11th April 2003,

[Zalba et al. 2004] Zalba B., Marin J.M., Cabeza L.F., Mehling H.: Free-cooling of buildings with phase change materials. International Journal of Refrigeration **27**, 839–849 (2004)

10 Appendix

Table 10.1. Nomenclature

	Si-units	Other common units		
A	m^2		area	
C	$J/kg \cdot K$, $J/m^3 \cdot K$, $J/mole \cdot K$		heat capacity	
c	C/mass, ...		specific heat capacity per mass, per volume, or per amount	
CR	$mg/m^2 yr$		corrosion rate	
d	m		thickness	
f			fraction of mass, volume, or amount	
H	J		enthalpy	
h	J/kg, J/m^3, J/mol		specific enthalpy per mass, per volume, or per amount	
k	$W/m^2 K$		overall heat transfer coefficient	
m	kg		mass	
mol	mol		amount of matter / number of molecules	
n			integer number	
P	$J/s = W$	kW	power	
p	Pa	bar ($=10^5$ Pa)	pressure	
Q	J	kWh	heat	
dQ/dt	W (=J/s)		heat flux	
dV/dt	m^3/s	liters/s	volume flow rate	
dq/dt	W/m^2		heat flux density	
r, R	m		radius	
R_{th}	K/W		thermal resistance	
S	J/K		entropy	
s	m		location of the phase front	
s^+			dimensionless ratio of radiuses	
T	°C		temperature	
ΔT	K		temperature difference	
t	s	min, hour	time	
U	J		internal energy	
u	m/s		flow speed	
V	m^3	liter ($=10^{-3}$ m^3)	volume	
x,y,z	m		Cartesian space coordinates	
α	$W/m^2 K$		convective heat transfer coefficient	

(Continued)

Table 10.1. (Continued)

	Si-units	Other common units	
α	m^2/s		thermal diffusivity
β	K/s	K/min	heating rate
β			dimensionless ratio of thermal resistances
η	N·s/m^2		viscosity
λ	W/mK		thermal conductivity
ρ	kg/m^3		density

Because of the frequent use of h as enthalpy, it would be confusing using also h as convective heat transfer coefficient. To avoid confusion, a different notation described in the following table is used throughout the book to describe heat transfer. Generally, the heat flux dQ/dt and the heat flux density dq/dt are related by dQ/dt = A · dq/dt.

Table 10.2. Notation for heat transfer.

effect	sketch	Notation often used	Notation here
Heat transfer dQ/dt through an area A by conduction		$dQ/dt = \dot{Q}$ $= A \cdot k \cdot \dfrac{\Delta T}{\Delta x}$ k = thermal conductivity	$dQ/dt = \dot{Q}$ $= A \cdot \lambda \cdot \dfrac{\Delta T}{\Delta x}$ λ = thermal conductivity
Heat transfer dQ/dt from an area A by convection		$dQ/dt = \dot{Q}$ $= A \cdot h_{conv} \cdot \Delta T$ h_{conv} = convection heat transfer coefficient	$dQ/dt = \dot{Q}$ $= A \cdot \alpha \cdot \Delta T$ α = convection heat transfer coefficient

effect	sketch	Notation often used	Notation here
Overall heat transfer dQ/dt through an area A, for example by convection on the surface and conduction inside	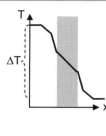	$dQ/dt = \dot{Q}$ $= A \cdot U \cdot \Delta T$ $1/U = 1/h_{cond} + \Delta x/\lambda$ $+ 1/h_{cond}$ U = overall heat transfer coefficient, or just U-factor R = 1/U overall thermal resistance, or just R-value with $dQ/dt = \dot{Q}$ $= A \cdot 1/R \cdot \Delta T$	$dQ/dt = \dot{Q}$ $= A \cdot k \cdot \Delta T$ $1/k = 1/\alpha + \Delta x/\lambda$ $+ 1/\alpha$ k = overall heat transfer coefficient

Table 10.3. Subscripts

a, amb	ambient
air	air
bp	boiling point
eff	effective
el	electric
f	fusion
i, j	integral numbers
in	ingoing
lm	logarithmic mean
max	maximum value
min	minimum value
mol	molar
mp	melting point
obj	object
out	outgoing
p	at constant pressure
pc	phase change
r	reference
s	sample
st	standard
th	thermal
V	at constant volume
wall	wall
water	water

Table 10.4. Superscripts.

a	air
amb	ambient
CV	control volume
HTF	heat transfer fluid
PCM	phase change material
w	water

Table 10.5. Abbreviations.

CFD	computational fluid dynamics
DHW	domestic hot water
DSC	differential scanning calorimetry
HEX	heat exchanger
HTF	heat transfer fluid
LHS	latent heat storage
mPCM	microencapsulated PCM
PC	Phase change
PEG	Polyethylen glycol
PCL	phase change liquid
PCM	phase change material
PCS	phase change slurries
ssPCM	shape stabilized PCM
TABS	thermally activated building systems
T-History	temperature history
TES	thermal energy storage

Table 10.6. Definitions within the text.

definition	section
thermal energy storage (TES)	1.1
sensible heat	1.1.1
heat capacity	1.1.1
latent heat	1.1.2
heat of fusion	1.1.2
phase change enthalpy	1.1.2
melting enthalpy	1.1.2
phase change material (PCM)	1.1.2
latent heat storage (LHS)	1.1.2
homogeneous evaporation-condensation	1.1.3
heterogeneous evaporation-condensation	1.1.3
heat of reaction	1.1.4
phase separation	2.1, 2.3.1
subcooling / supercooling	2.1, 2.3.2
cycling stability	2.1
eutectic water-salt solutions	2.2.2
eutectic composition	2.2.2
salt hydrate	2.2.2
paraffin	2.2.2
fatty acid	2.2.2
polyethylene glycol	2.2.2
clathrate	2.2.2
gas hydrate	2.3.1
congruent melting	2.3.1
decomposition / phase separation	2.3.1
semicongruent melting	2.3.1
incongruent melting	2.3.1
artificial mixing	2.3.1
gelling	2.3.1
thickening	2.3.1
homogeneous nucleation	2.3.2
heterogeneous nucleation	2.3.2
nucleator	2.3.2
cold finger	2.3.2
macroencapsulation	2.3.3
microencapsulation	2.3.3
material	3.1
material property	3.1

composite material	3.1
object property	3.1
calorimetric formula	3.2.1
calorimetric methods	3.2.2
calorimeter	3.2.2
interval mode	3.2.2
step mode	3.2.2
dynamic mode	3.2.2
hysteresis	3.2.3
apparent hysteresis	3.2.3
Differential scanning calorimeter (DSC)	3.2.4.1
hf-DSC	3.2.4.1
hot-wire method	3.4
finite difference method	4.2.1
explicit method	4.2.1
enthalpy method	4.2.1
mushy zone	4.2.3
dynamic storage	5.2
static storage	5.2
active storage	5.2
passive storage	5.2
heat exchanger type	5.2
direct contact type	5.2
module type	5.2
slurry type	5.2
sensible type	5.2
useful stored heat	5.4.1.3
porous medium model	5.4.3.3
phase change slurry (PCS)	5.5.1
phase change liquid (PCL)	5.5.1
cascade storage	6.3
human comfort requirements	8.1.1
operative temperature	9.1.1
air-conditioning	9.1.1
free cooling	9.1.3.2
coefficient of performance, COP	9.1.4.1
thermally activated building systems, TABS	9.2.4
ice bank	9.2.5.1
ice on coil	9.2.5.1
ice-harvesting system	9.2.5.3
ice maker, ice harvester, ice generator	9.2.5.3
secondary refrigerant	9.2.5.6

11 Index

A
active, 141
air-conditioning, 219, 248, 257
alcane, 20, 43
aluminum foams, 40
analytical model, 150
analytical solution, 106, 119
apparent hysteresis, 65
artificial mixing, 32

B
binary ice, 170
blinds, 243
boundary conditions, 137
building components, 241
building materials, 221, 224, 235, 240
building simulation software, 131
building ventilation, 164

C
calibration, 64, 76
calorimeter, 60
calorimetric formula, 59
calorimetric methods, 60
calorimetry, 58
cascade storage, 185
ceiling, 241, 245, 256
CFD, 131
characteristic numbers, 176
chemical stability, 12
clathrate, 25, 270
coacervation, 38
coating, 208
coefficient of performance (COP), 230
cold finger technique, 36
cold sources, 227
compatibility, 12, 15, 97, 101
composite, 39
composite material, 39, 41, 43, 57, 209
computational fluid dynamics, 131
concrete, 238, 255
congruent, 26
construction, 12
convection, 154, 158
cooling experiment, 133
cooling vest, 210
corrosion, 15, 97
corrosivity, 12
crosslink, 38
crystallization, 68, 127
cycles, 11
cycling, 11
cycling stability, 11, 95
cycling tests, 95
cylindrical geometry, 113

D
decomposition, 28
design options, 137
differential equations, 120
differential scanning calorimetry (DSC), 69
dimple sheets, 49
direct contact type, 158, 263, 288
domestic hot water, 175, 226
Dulong-Petit, 8
dynamic, 141
dynamic method, 93
dynamic mode, 61, 74

E
economic requirements, 13
electronic equipment, 201
empirical model, 161
emulsifier, 171
emulsion, 38, 171
emulsion slurry, 171
encapsulation, 37, 48
enthalpy, 3, 59, 123
enthalpy calibration, 75
enthalpy method, 124, 168
erythritol, 23
eutectic composition, 15
eutectic mixture, 29, 128
eutectic point, 30
eutectic water-salt solution, 15, 43
expanded graphite, 47
experimental data, 132
explicit method, 123

F
fatty acid, 22
fiber, 45, 208
finite difference method, 121
fins, 146, 149, 158
first law of thermodynamics, 59
flat plates, 163, 164
floor, 247
floor heating, 280
forced convection, 112
free convection, 113
free cooling, 226
functional textiles, 207

G

Galisol principle, 161
gas hydrate, 25
gelatine, 38
gelling, 33
Glaubersalt, 31
granulate, 44
graphite matrix, 45, 47
Gütegemeinschaft PCM e.V., 42

H

heat capacity, 2, 59, 66, 89, 123
heat capacity calibration, 72
heat capacity method, 168
heat exchanger, 132, 148
heat exchanger type, 146, 258, 284
heat exchanging calorimeters, 69
heat flux density, 107
heat flux DSC, 69
heat of fusion, 3
heat of reaction, 5
heat sources, 231
heat stored, 61, 88
heat transfer, 105, 137, 233
heat transfer calculation, 144, 145, 149, 161, 164, 172, 176
heat transfer coefficient, 110, 234
heat transfer medium, 85
heating rate, 71, 75
heterogeneous evaporation-condensation, 4
heterogeneous nucleation, 36
hf-DSC, 69
homogeneous, 58
homogeneous evaporation-condensation, 4
homogeneous nucleation, 36
hot water heat storage, 2
hot-wire method, 93
human body, 205
human comfort requirements, 206, 218
hybrid storage, 163
hysteresis, 65, 67

I

ice bank, 259
ice generator, 263
ice harvester, 263
ice slurry, 170
ice-harvesting systems, 263
icemaker, 263
ice-on-coil, 258
implicit method, 123
impregnation, 40
incompressible, 59
incongruent, 26, 130
incongruent melting, 32
inhomogeneities, 63
inorganic materials, 15
insulated environment, 143
integration concepts, 182, 271
intercomparison test, 69
internal energy, 59
internal heat transfer surfaces, 146
interval mode, 60
isothermal, 71, 88

L

latent heat, 3
latent heat storage, 3
latent heat storage material, 3
lifetime, 95

M

macroencapsulated PCM, 162, 207
macroencapsulation, 37, 49
material, 57
material classes, 13
material property, 57, 58, 91
m-DSC, 80
mechanical stability, 12
medical applications, 200, 214
melting, 2
melting enthalpy, 3, 15, 76
melting range, 63
melting temperature, 3, 62
microencapsulated paraffin, 236
microencapsulated PCM, 169, 207
microencapsulated PCM slurry, 171
microencapsulation, 37, 51
mixing calorimeter, 89
mixture, 19, 43, 129
module type, 162, 163, 258, 286
mushy zone, 130

N

natural ice, 264
natural ventilation, 242
node, 121, 154, 164
normalization, 88
nucleator, 36
nucleus, 36
numerical model, 120, 132, 153

O

object, 58
object property, 58
operative temperature, 219, 241
organic materials, 15

P

packed bed, 164, 177
paraffin, 20, 43

paraffin-graphite composite, 250
passive, 141
passive system, 242
PCM, 95
PCM module, 162
PCM-composite material, 95
PCM-graphite composite, 41, 45
PCM-graphite compound, 48
PCM-graphite matrix, 47, 154
PCM-object, 84, 95
Peltier element, 36
peritectic temperature, 31
peritectic transformation, 30
phase change enthalpy, 3, 11, 66
phase change liquid (PCL), 169
phase change material, 3
phase change slurry (PCS), 169, 266
phase change temperature, 3, 11, 66, 223
phase change temperature range, 223
phase diagram, 128
phase front, 152
phase separation, 11, 26, 28, 159
physical requirements, 11
pizza heater, 199
plaster, 236, 255
plasterboard, 236
plastics, 101
pocket heater, 210
polycondensation, 38
polyethylene glycol, 24
polymerization, 38
porous medium, 168
porous medium model, 164
powder, 44, 45
power, 12, 88, 146, 149
power compensated DSC, 90
power compensating calorimeters, 69

R

reaction equilibrium, 63
recyclability, 13
reference, 89
reproducibility, 95
reproducible phase change, 11
Richardson, 9

S

safety, 12
salt, 18
salt hydrate, 16, 43
sample, 58
seasonal heat storage, 226
secondary nucleation, 36
secondary refrigerant, 267

semi congruent, 30, 130
semi congruent melting, 31, 32
semi-infinite layer, 106
sensible liquid type, 174
sensitivity, 71, 72, 74
shape-stabilized PCM, 39, 240, 281
shape-stabilized PCM slurry, 172
size, 58
slurry, 169
slurry type, 169, 266, 269
snow, 264
solidification, 2
space cooling, 227, 234
space cooling applications, 225
space heating, 217, 226, 231, 233, 273
specific heat capacity, 2
spherical capsules, 164
spherical geometry, 113
standard, 41, 58, 69, 97
standard material, 74, 76
static, 141
static storages, 258
stationary method, 92
Stefan problem, 107, 112, 131
steps mode, 61, 78
storage container, 191
storage density, 146
storage medium, 2
storage module, 153
storage type, 141, 146
stored heat, 2, 59, 67, 79, 95
stratification, 173, 176
structure, 58
subcooling, 11, 34, 68, 81, 126, 133, 168, 210
sugar alcohol, 23
supercooling, 12
system, 181, 188

T

technical requirements, 12
temperature control, 142, 143, 221
temperature equilibrium, 65
temperature modulated DSC, 80
test chamber, 88
thermal conductivity, 12, 40, 58, 64, 91, 96
thermal diffusivity, 93
thermal energy storage, 1
thermal mass, 221
thermal resistance, 69, 109
Thermally Activated Building Systems, TABS, 255
thermally conducting structure, 45
thermodynamic equilibrium, 61, 63, 64

thermodynamic property, 132
thickening, 33
T-History method, 80
tightness, 12
total heat stored, 149
transport box, 143
transport container, 191, 197
Trombe wall, 239
Trouton, 9
U
useful stored heat, 158

V
vacuum super insulation, 144
vapor pressure, 12
ventilation, 249
viscosity, 33
volume change, 12
volume element, 121
W
wall, 246, 274
water, 35, 89, 126
X
xylitol, 23